20.95
80U

LONDON MATHEMATICAL SOCIETY LECTURE NOTE SERIES

Managing Editor: Professor J.W.S. Cassels, Department of Pure Mathematics and Mathematical Statistics, 16 Mill Lane, Cambridge CB2 1SB, England

The books in the series listed below are available from booksellers, or, in case of difficulty, from Cambridge University Press.

4 Algebraic topology, J.F.ADAMS
5 Commutative algebra, J.T.KNIGHT
11 New developments in topology, G.SEGAL (ed)
12 Symposium on complex analysis, J.CLUNIE & W.K.HAYMAN (eds)
13 Combinatorics, T.P.McDONOUGH & V.C.MAVRON (eds)
16 Topics in finite groups, T.M.GAGEN
17 Differential germs and catastrophes, Th.BROCKER & L.LANDER
18 A geometric approach to homology theory, S.BUONCRISTIANO, C.P.ROURKE & B.J.SANDERSON
20 Sheaf theory, B.R.TENNISON
21 Automatic continuity of linear operators, A.M.SINCLAIR
23 Parallelisms of complete designs, P.J.CAMERON
25 Lie groups and compact groups, J.F.PRICE
26 Transformation groups, C.KOSNIOWSKI (ed)
27 Skew field constructions, P.M.COHN
29 Pontryagin duality and the structure of LCA groups, S.A.MORRIS
30 Interaction models, N.L.BIGGS
31 Continuous crossed products and type III von Neumann algebras,A.VAN DAELE
32 Uniform algebras and Jensen measures, T.W.GAMELIN
34 Representation theory of Lie groups, M.F. ATIYAH et al.
35 Trace ideals and their applications, B.SIMON
36 Homological group theory, C.T.C.WALL (ed)
37 Partially ordered rings and semi-algebraic geometry, G.W.BRUMFIEL
38 Surveys in combinatorics, B.BOLLOBAS (ed)
39 Affine sets and affine groups, D.G.NORTHCOTT
40 Introduction to Hp spaces, P.J.KOOSIS
41 Theory and applications of Hopf bifurcation, B.D.HASSARD, N.D.KAZARINOFF & Y-H.WAN
42 Topics in the theory of group presentations, D.L.JOHNSON
43 Graphs, codes and designs, P.J.CAMERON & J.H.VAN LINT
44 Z/2-homotopy theory, M.C.CRABB
45 Recursion theory: its generalisations and applications, F.R.DRAKE & S.S.WAINER (eds)
46 p-adic analysis: a short course on recent work, N.KOBLITZ
47 Coding the Universe, A.BELLER, R.JENSEN & P.WELCH
48 Low-dimensional topology, R.BROWN & T.L.THICKSTUN (eds)
49 Finite geometries and designs,P.CAMERON, J.W.P.HIRSCHFELD & D.R.HUGHES (eds)
50 Commutator calculus and groups of homotopy classes, H.J.BAUES
51 Synthetic differential geometry, A.KOCK
52 Combinatorics, H.N.V.TEMPERLEY (ed)
54 Markov process and related problems of analysis, E.B.DYNKIN
55 Ordered permutation groups, A.M.W.GLASS
56 Journêes arithmêtiques, J.V.ARMITAGE (ed)
57 Techniques of geometric topology, R.A.FENN
58 Singularities of smooth functions and maps, J.A.MARTINET
59 Applicable differential geometry, M.CRAMPIN & F.A.E.PIRANI
60 Integrable systems, S.P.NOVIKOV et al
61 The core model, A.DODD
62 Economics for mathematicians, J.W.S.CASSELS
63 Continuous semigroups in Banach algebras, A.M.SINCLAIR
64 Basic concepts of enriched category theory, G.M.KELLY
65 Several complex variables and complex manifolds I, M.J.FIELD

66 Several complex variables and complex manifolds II, M.J.FIELD
67 Classification problems in ergodic theory, W.PARRY & S.TUNCEL
68 Complex algebraic surfaces, A.BEAUVILLE
69 Representation theory, I.M.GELFAND _et al._
70 Stochastic differential equations on manifolds, K.D.ELWORTHY
71 Groups - St Andrews 1981, C.M.CAMPBELL & E.F.ROBERTSON (eds)
72 Commutative algebra: Durham 1981, R.Y.SHARP (ed)
73 Riemann surfaces: a view towards several complex variables,A.T.HUCKLEBE
74 Symmetric designs: an algebraic approach, E.S.LANDER
75 New geometric splittings of classical knots, L.SIEBENMANN & F.BONAHON
76 Spectral theory of linear differential operators and comparison algebras, H.O.CORDES
77 Isolated singular points on complete intersections, E.J.N.LOOIJENGA
78 A primer on Riemann surfaces, A.F.BEARDON
79 Probability, statistics and analysis, J.F.C.KINGMAN & G.E.H.REUTER (eds
80 Introduction to the representation theory of compact and locally compact groups, A.ROBERT
81 Skew fields, P.K.DRAXL
82 Surveys in combinatorics, E.K.LLOYD (ed)
83 Homogeneous structures on Riemannian manifolds, F.TRICERRI & L.VANHECKE
84 Finite group algebras and their modules, P.LANDROCK
85 Solitons, P.G.DRAZIN
86 Topological topics, I.M.JAMES (ed)
87 Surveys in set theory, A.R.D.MATHIAS (ed)
88 FPF ring theory, C.FAITH & S.PAGE
89 An F-space sampler, N.J.KALTON, N.T.PECK & J.W.ROBERTS
90 Polytopes and symmetry, S.A.ROBERTSON
91 Classgroups of group rings, M.J.TAYLOR
92 Representation of rings over skew fields, A.H.SCHOFIELD
93 Aspects of topology, I.M.JAMES & E.H.KRONHEIMER (eds)
94 Representations of general linear groups, G.D.JAMES
95 Low-dimensional topology 1982, R.A.FENN (ed)
96 Diophantine equations over function fields, R.C.MASON
97 Varieties of constructive mathematics, D.S.BRIDGES & F.RICHMAN
98 Localization in Noetherian rings, A.V.JATEGAONKAR
99 Methods of differential geometry in algebraic topology, M.KAROUBI & C.LERUSTE
100 Stopping time techniques for analysts and probabilists, L.EGGHE
101 Groups and geometry, ROGER C.LYNDON
102 Topology of the automorphism group of a free group, S.M.GERSTEN
103 Surveys in combinatorics 1985, I.ANDERSEN (ed)
104 Elliptic structures on 3-manifolds, C.B.THOMAS
105 A local spectral theory for closed operators, I.ERDELYI & WANG SHENGWAN
106 Syzygies, E.G.EVANS & P.GRIFFITH
107 Compactification of Siegel moduli schemes, C-L.CHAI
108 Some topics in graph theory, H.P.YAP
109 Diophantine analysis, J.LOXTON & A.VAN DER POORTEN (eds)
110 An introduction to surreal numbers, H.GONSHOR
111 Analytical and geometric aspects of hyperbolic space, D.B.A.EPSTEIN (ec
112 Low dimensional topology and Kleinian groups, D.B.A.EPSTEIN (ed)
113 Lectures on the asymptotic theory of ideals, D.REES
114 Lectures on Bochner-Riesz means, K.M.DAVIS & Y-C.CHANG
115 An introduction to independence for analysts, H.G.DALES & W.H.WOODIN
116 Representations of algebras, P.J.WEBB (ed)
117 Homotopy theory, E.REES & J.D.S.JONES (eds)
118 Skew linear groups, M.SHIRVANI & B.WEHRFRITZ

London Mathematical Society Lecture Note Series. 97

Varieties of Constructive Mathematics

DOUGLAS BRIDGES
University of Buckingham

FRED RICHMAN
New Mexico State University

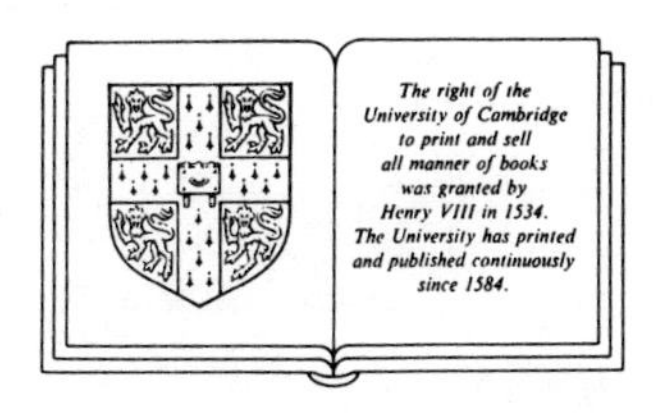

CAMBRIDGE UNIVERSITY PRESS
Cambridge
New York New Rochelle
Melbourne Sydney

Published by the Press Syndicate of the University of Cambridge
The Pitt Building, Trumpington Street, Cambridge CB2 1RP
32 East 57th Street, New York, NY 10022, USA
10 Stamford Road, Oakleigh, Melbourne 3166, Australia

First published 1987
Reprinted 1988

Printed in Great Britain at the University Press, Cambridge

Library of Congress cataloguing in publication data

Bridges, D. S. (Douglas S.), 1945 -
Varieties of constructive mathematics.
(London Mathematical Society lecture note series ;97)
1. Constructive mathematics. I. Richman, Fred.
II. Title III. Series
QA9.56.B75 1986 511.3 85-26904

British Library cataloguing in publication data

Bridges, Douglas
Varieties of constructive mathematics.---
(London Mathematical Society lecture note series;
ISSN 0076-0552; v. 97)
1. Logic, Symbolic and mathematical
I. Title II. Richman, Fred III. Series
511.3 BC135

ISBN 0 521 31802 5

Preface

In Hilary Term, 1981, Douglas Bridges gave a course of lectures on Intuitionism and constructive mathematics in the Mathematical Institute of Oxford University. Shortly afterwards, he invited Fred Richman to join in the writing of a book based (as it turns out, rather loosely) on those lectures. The book now lies in front of the reader, as an introduction to the spirit and practice of modern constructive mathematics.

There are several excellent works - such as Beeson's *Foundations of Constructive Mathematics* and Dummett's *Elements of Intuitionism* - on the logical and philosophical foundations of constructive mathematics; and there are others, such as Bishop's seminal treatise *Foundations of Constructive Analysis*, dealing with the detailed development of major portions of mathematics within a constructive framework. The present book is intended to land between those two positions: specifically, we hope that, with a minimum of philosophy and formal logic, and without requiring of the reader too great an investment of time and effort over technical details, it will leave him with a clear conception of the problems and methods of the three most important varieties of modern constructive mathematics.

Since classical mathematics, as practised by all but a tiny minority of mathematicians, appears to offer a much less arduous route to discovery, and a far greater catalogue of successes, than its constructive counterpart, one may well ask: Why should anyone, other than a devotee of a constructivist philosophy, be interested in learning about constructive mathematics? We believe there are several reasons why one might be so interested.

First, there is the richer structure of constructive mathematics that flows from the deepened meaning of existence. In the classical interpretation, an object exists if its non-existence is contradictory. There is a clear distinction between this meaning of existence and the constructive, algorithmic one, under which an object exists only if we can construct it, at least in principle. As Bishop has said, such 'meaningful distinctions deserve to be maintained'.

Second, there is the unexpected role played by intuitionistic logic and constructive methods in topos theory. Any theorem in constructive mathematics may be interpreted as a classical theorem about a topos; if the topos is suitably chosen, then the theorem may have classical interest outside of topos theory. For example, constructive theorems about diagonalizing matrices over the reals may be interpreted as classical theorems about diagonalizing matrices over the ring of continuous functions on a compact metric space.

Finally, there is the possibility of applications of constructive mathematics to areas such as numerical mathematics, physics, and computer science. For computer science, we refer the reader to Martin-Löf's paper *Constructive mathematics and computer programming* (in: *Logic, Methodology and the Philosophy of Science* VI, North-Holland, 1982), and to the recent book *Implementing Mathematics with the Nuprl Proof Development System*, by Constable et al. (Prentice-Hall, 1986).

We now outline the contents of our book. In Chapter 1 we introduce the three varieties of constructive mathematics with which we shall be concerned: Bishop's constructive mathematics, Brouwer's intuitionistic mathematics, and the constructive recursive mathematics of the Russian school of Markov; we also construct the real line $\mathbb{R}$ and examine its basic properties.

In Chapter 2 we discuss, within Bishop's mathematics, a range of topics in analysis. These topics are chosen to illustrate distinctive features of the practice of constructive mathematics, such as the splitting of one classical theorem (for example, Baire's theorem) into several inequivalent constructive ones, and the investigation of important notions, like locatedness, with little or no classical significance. A similar approach to constructive algebra is found in Chapter 4, which culminates in a constructive treatment of the Hilbert basis theorem. Chapter 4 also offers some comparisons between constructive algebra and classical recursive algebra.

Chapter 3 contains the essentials of constructive recursive mathematics, and is based on Richman's axiomatic approach to Church's thesis. This approach greatly simplifies the presentation of mathematics within the recursive framework, and leads to particularly perspicuous proofs of Specker's theorem and the existence of a compact subset of $\mathbb{R}$ that is not Lebesgue measurable.

Chapter 5 deals with intuitionistic mathematics, by formulating axioms which capture the mathematical essence of Brouwer's approach. The elements of intuitionistic mathematics are developed, including Brouwer's famous theorem that every real-valued function on a compact interval is uniformly continuous.

In Chapter 6 the three varieties of constructive mathematics are compared by an examination of the status within each of the classical proposition

if f *is a uniformly continuous mapping of* $[0,1]$ *into the positive real line, then the infimum of* f *is positive.*

The final chapter deals with intuitionistic logic and topos models, mainly through the presentation of a few examples. We treat this material somewhat superficially, not wishing to involve the reader in a detailed development of the logic and category theory necessary for a fully rigorous treatment.

A remark about our style of references is in order here. With few exceptions, references are found in the notes at the end of each chapter, in which case they are normally described in detail. There are certain works to which we refer so often, or which are indispensable to the practising constructive mathematician, that we chose to give them special names, printed in bold face; these are

Beeson	Michael J. Beeson, *Foundations of Constructive Mathematics* (Springer, 1985)
Bishop	Errett Bishop, *Foundations of Constructive Analysis*, (McGraw-Hill, 1967)
Bishop-Bridges	Errett Bishop and Douglas Bridges, *Constructive Analysis* (Springer, 1985)
Brouwer	*Brouwer's Cambridge lectures on intuitionism*, (Dirk Van Dalen ed., Cambridge University Press, 1981)
Dummett	Michael Dummett, *Elements of Intuitionism* (Oxford University Press, 1977)

Kushner	B.A. Kushner, *Lectures on Constructive Mathematical Analysis* (American Mathematical Society, 1985)
MRR	Ray Mines, Fred Richman, Wim Ruitenberg, *A Course on Constructive Algebra* (Springer, forthcoming)
Springer 873	*Constructive Mathematics* (Fred Richman ed., Springer Lecture Notes in Mathematics 873)

During the writing of this book, we received support from the University of Buckingham and from New Mexico State University. As usual, we have had many hours of stimulating discussion with Bill Julian and Ray Mines. Bill and Nancy Julian kindly provided hospitality when Bridges visited New Mexico to work with Richman. David Tranah, of Cambridge University Press, has been unbelievably patient with us, as deadline after deadline has passed by.

Perhaps the most long-suffering have been our families, who have had to share the birth pangs of this book over an unusually long labour; we dedicate this work to them.

Douglas S. Bridges
Fred Richman

Buckingham, England
Las Cruces, New Mexico

Contents

Chapter 1. Foundations of Constructive Mathematics

In which the reader is introduced to the three varieties of constructive mathematics that will be studied in detail in subsequent chapters; the framework of Bishop's constructive mathematics is erected; and the elementary theory of the real numbers is developed.

1. Existence and omniscience

We engage in constructive mathematics from a desire to clarify the meaning of mathematical terminology and practice – in particular, the meaning of existence in a mathematical context. The classical mathematician, with the freedom of methodology advocated by Hilbert, perceives an object x to exist if he can prove the impossibility of its nonexistence; the constructive mathematician must be presented with an algorithm that constructs the object x before he will recognize that x exists.

What do we mean by an *algorithm*? We may think of an algorithm as a specification of a step-by-step computation, such as a program in some computer language, which can be performed, at least in principle, by a human being or a computer in a finite period of time; moreover, the passage from one step to another should be deterministic. Note that we say 'performed, at least in principle', for it is possible for an algorithm to require an amount of time greater than the age of the universe for its complete execution. We are not concerned here with questions of complexity or efficiency.

In Bishop's constructive mathematics (**BISH**), and in Brouwer's intuitionism (**INT**), the notion of an algorithm, or finite routine, is taken as primitive. Russian constructivism (**RUSS**), on the other hand, operates within a fixed programming language, and an algorithm is a sequence of symbols in that language.

While BISH is only one of the three varieties of constructive mathematics that we shall consider, it is the one to which we shall devote the most attention. There are at least two reasons for this. First, BISH is consistent with **CLASS**, the classical mathematics practised by most mathematicians today. Every proposition P in BISH has an immediate interpretation in CLASS, and a proof of P in BISH is also a proof of P in CLASS. This is not true in our other two varieties: Russian constructivism (RUSS) and Brouwer's intuitionism (INT). For example, in both RUSS and INT it is proved that every real valued function on the interval [0,1] is pointwise continuous.

A second reason for paying particular attention to BISH is that every proof of a proposition P in BISH is a proof of P in RUSS and in INT. Indeed, the last two varieties may be regarded as extensions of BISH. In RUSS, the main principle adjoined to BISH is a form of Church's thesis that all sequences of natural numbers are recursive. In INT, two principles are added which ensure strong continuity properties of arbitrary real-valued functions on intervals of the line.

From a philosophical point of view, there is more to RUSS and INT than the mere adjunction of certain principles to BISH. In RUSS, every mathematical object is, ultimately, a natural number: constructions take place within a fixed formal system, functions are Gödel numbers of the algorithms that compute them, and so on. On the other hand, INT is based on Brouwer's intuitionistic philosophy, including an analysis of the notion of an infinitely proceeding, or free choice, sequence. For the most part we shall ignore the philosophical aspects by abstracting in each case the features essential for the development of the associated mathematics; we are writing for mathematicians rather than for philosophers or logicians.

The essential difference between BISH and CLASS is illustrated by considering the simplest kind of statement concerning existence in an infinite context. A **binary sequence** is a finite routine that assigns to each positive integer an element of $\{0,1\}$. Let a be a binary sequence, and consider the statements

$$
\begin{aligned}
P(a) &: \quad a_n = 1 \text{ for some } n, \\
\neg P(a) &: \quad a_n = 0 \text{ for all } n, \\
P(a) \vee \neg P(a) &: \quad \text{Either } P(a) \text{ or } \neg P(a),
\end{aligned}
$$

$$\forall\alpha(P(\alpha) \vee \neg P(\alpha)) : \text{ For all } \alpha, \text{ either } P(\alpha) \text{ or } \neg P(\alpha).$$

Note that $\neg P(\alpha)$ is the denial of $P(\alpha)$. A constructive proof of $P(\alpha) \vee \neg P(\alpha)$ must provide a finite routine which either shows that $a_n = 0$ for all n, or computes a positive integer n with $a_n = 1$. (Note the constructive meaning of disjunction, which follows from the constructive interpretation of existence: $P_1 \vee P_2$ holds if and only if there exists i such that P_i; to prove $P_1 \vee P_2$, it is not enough to show that P_1 and P_2 cannot both be false.) In particular, if α is the binary sequence defined by setting

$$\begin{aligned} a_n &= 0 \quad \text{if } x^{m+2} + y^{m+2} \neq z^{m+2} \text{ for all} \\ &\qquad \text{positive integers } x,y,z,m \leq n, \\ &= 1 \quad \text{otherwise,} \end{aligned}$$

then a constructive proof of $P(\alpha) \vee \neg P(\alpha)$ would give a method for deciding the Fermat conjecture, $\neg P(\alpha)$, by providing a construction that either establishes the Fermat conjecture or produces an explicit counterexample to it; unless we have such a construction, we are not entitled to assert $P(\alpha) \vee \neg P(\alpha)$ within BISH. We say that α is a **Brouwerian example** of a binary sequence for which $P(\alpha) \vee \neg P(\alpha)$ does not hold, or that α is a **Brouwerian counterexample** to the statement $\forall\alpha(P(\alpha) \vee \neg P(\alpha))$. Note that a Brouwerian counterexample is not an counterexample in the usual sense; it is *evidence* that a statement does not admit a constructive proof.

The use of Fermat's conjecture in the above example is not essential. If Fermat's conjecture were resolved tomorrow, we could then assert $P(\alpha) \vee \neg P(\alpha)$; however, by referring to another open problem, such as the Riemann hypothesis or the Goldbach conjecture, we could construct another binary sequence α for which we could not assert $P(\alpha) \vee \neg P(\alpha)$

More generally, a **Brouwerian counterexample** to an assertion A is a proof that A implies some principle that is unacceptable, or at least highly dubious, in the context of constructive mathematics. The two most popular principles were introduced by Brouwer; we present them here under the names given them by Bishop. The first is simply the assertion $\forall\alpha(P(\alpha) \vee \neg P(\alpha))$; the second is a little subtler.

LPO **The limited principle of omniscience:** *If* (a_n) *is a binary sequence, then either there exists* n *such that* $a_n = 1$, *or else* $a_n = 0$ *for each* n.

LLPO **The lesser limited principle of omniscience:** *If* (a_n) *is a binary sequence containing at most one* 1, *then either* $a_{2n} = 0$ *for each* n, *or else* $a_{2n+1} = 0$ *for each* n.

A constructive proof of LPO would provide a finite decision procedure for a vast number of unsolved problems in mathematics, including the conjectures of Fermat, Riemann, and Goldbach. The unlikelihood of finding a method of such power and scope is an argument for excluding LPO from constructive mathematics, at least provisionally; were such a method found, our ideas concerning constructive mathematics would have to be drastically revised, especially as LPO and LLPO are provably false within INT and RUSS.

The **law of excluded middle**, which asserts that $P \vee \neg P$ holds for any statement P, implies LPO, and so is rejected in constructive mathematics.

Given a real number r, for each positive integer n compute a rational number r_n such that $|r - r_n| < 1/n$. If $|r_n| \leq 1/n$, set $a_n \equiv 0$, and if $|r_n| > 1/n$, set $a_n \equiv 1$. Then the resulting binary sequence $a \equiv (a_1, a_2, \ldots)$ has the property that

$$(*) \qquad \begin{array}{l} |r| > 0 \text{ if and only if } \exists n(a_n = 1), \text{ and} \\ r = 0 \text{ if and only if } \forall n(a_n = 0). \end{array}$$

Conversely, to each binary sequence a there corresponds a real number $r \equiv \sum_{n=1}^{\infty} 2^{-n} a_n$ such that (*) obtains. Thus the limited principle of omniscience is equivalent to the assertion that each real number is either 0 or different from 0. Another assertion equivalent to LPO is the **trichotomy law:** $[-1,1] = [-1,0) \cup \{0\} \cup (0,1]$.

The lesser limited principle of omniscience says that for any binary sequence, either the first nonzero term, if it exists, has even index, or the first nonzero term, if it exists, has odd index. This is equivalent to the assertions

$$[-1,1] = [-1,0] \cup [0,1],$$

and

for each real number r, either $r \leq 0$ or $r \geq 0$.

Another assertion equivalent to LLPO is that if the product of two real numbers is 0, then one or the other of them is 0.

Many other familiar results of CLASS are equivalent to some principle that is rejected in BISH. The intermediate value theorem for uniformly continuous functions is equivalent to LLPO, while LPO is equivalent to the assertion that every real number is either rational or irrational. The law of excluded middle is a consequence of the statement that every ideal in the ring of integers is finitely generated.

It is helpful to formulate omniscience principles in the notation of the (intuitionistic) propositional calculus, in terms of the simplest kind of assertion of existence in an infinite context. Let us call an assertion A **simply existential** if we can construct a binary sequence α such that A holds if and only if there exists n such that $\alpha_n = 1$. For arbitrary simply existential statements A and B, the principles LPO and LLPO take the forms:

$$\text{LPO:} \quad A \vee \neg A,$$

$$\text{LLPO:} \quad \neg(A \wedge B) \Leftrightarrow \neg A \vee \neg B.$$

It is easily seen that if A and B are simply existential, then so is $A \wedge B$. Thus LLPO can be extended, by induction on n, to read

$$\neg(A_1 \wedge A_2 \wedge \cdots \wedge A_n) \Leftrightarrow \neg A_1 \vee \neg A_2 \vee \cdots \vee \neg A_n.$$

Two other omniscience principles, stated for an arbitrary simply existential statement A, are the **weak limited principle of omniscience**

$$\text{WLPO:} \quad \neg A \vee \neg\neg A,$$

and **Markov's Principle**

$$\text{MP:} \quad \neg\neg A \Leftrightarrow A.$$

These two principles were also rejected by Brouwer, and are commonly used by constructive mathematicians for counterexamples. It is not difficult to show that LPO implies WLPO, and that WLPO implies LLPO. These three principles are false in RUSS, unlike Markov's principle, which is used freely by the practitioners of that school.

The reader should not be misled into thinking that constructive mathematics is largely a criticism of classical mathematics by means of Brouwerian counterexamples, although such counterexamples do play an important role in establishing the direction of constructive research. The main task of the constructive mathematician is the positive development of mathematics, along lines suggested by familiar classical theories, or in altogether new directions. A proof in BISH of a long-established classical theorem may contain computational information, of interest to a classical mathematician, that is not yielded by the standard proofs in CLASS. Moreover, proofs in BISH may provide more insight than their classical predecessors, and are often more natural and elegant. It is to this positive development that we now turn.

2. Basic constructions

The foundation stones of constructive mathematics are the positive integers:

> *The primary concern of mathematics is number, and this means the positive integers... Everything attaches itself to number, and every mathematical statement ultimately expresses the fact that if we perform certain computations within the set of positive integers, we shall obtain certain results.* [**Bishop**, pp. 2–3]

We take as basic our intuition of the positive integers and their arithmetical properties, including induction; so we make no attempt to explain them, for example, in terms of an axiomatic theory. We denote the set of positive integers by $\mathbb{N}^+$, and the set of nonnegative integers (natural numbers) by $\mathbb{N}$.

With the positive integers at the foundation, each successive level of the building is erected upon the levels below: at any stage, objects may be constructed using those objects which are, or could have been, constructed at previous stages. The new constructs may be ordered pairs of previously constructed objects, sets of such objects, functions associating objects of one set with objects of another, and so on.

To define a **set** S we must (i) explain how to construct elements (members) of S using objects that have been, or could have been, constructed prior to S, and (ii) describe what it means for two elements of S to be equal. The **equality** on a set is an essential part of its description, and must satisfy the defining properties of an **equivalence relation:**

(2.1) (i) $x = x$;

(ii) if $x = y$, then $y = x$;

(iii) if $x = y$ and $y = z$, then $x = z$.

We write $x \in S$ to mean that x is an element of S.

A property P which is applicable to the elements of a set A determines a **subset** S of A which is denoted by $\{x \in A : P(x)\}$: if x is an element of A, then $x \in S$ if and only if $P(x)$. Here, as elsewhere, we are only interested in a property $P(x)$ that is **extensional**, in the sense that for all x, x' in A with $x = x'$, $P(x)$ if and only if $P(x')$. Informally, to say that $P(x)$ is extensional means that it does not depend on the particular description by which x is given to us.

If S is a subset of A, and $x \in A$, we write $x \notin S$ to mean that $x \in S$ is impossible; and we denote by $A \backslash S$ the set $\{x \in A : x \notin S\}$.

The set of all ordered pairs of elements of a set S, taken with the equality

$$(x,y) = (x',y') \text{ if and only if } x = x' \text{ and } y = y',$$

is called the **Cartesian product** of S with itself, and is written $S \times S$. A subset of $S \times S$, or, equivalently, a property applicable to elements of $S \times S$, is called a **binary relation** on S.

Many sets also come equipped with a binary relation $\neq$, called an **inequality**, satisfying the defining properties:

(2.2) (i) if $x \neq y$, then $y \neq x$;

(ii) $x \neq x$ is impossible.

Note that, by our implicit assumption of extensionality, an inequality satisfies the condition

$$\text{if } x \neq y \text{ and } y = z, \text{ then } x \neq z.$$

An inequality that also satisfies the condition

$$\text{if } x \neq y, \text{ then for all } z \text{ either } x \neq z \text{ or } y \neq z$$

is called an **apartness**. An apartness is **tight** if $x = y$ whenever $x \neq y$ is impossible. A set with an inequality is **discrete** if, for any two elements x and y, either $x = y$ or $x \neq y$. The **denial inequality**, obtained by setting $x \neq y$ if $x = y$ is impossible, need not be an apartness. If a set is discrete under the denial inequality, we often simply say that it is discrete.

The set $\mathbb{N}$ of natural numbers is discrete. If n is a positive integer, then the discrete set $\mathbb{Z}_n$ is constructed by stipulating that its elements are integers, and that $x = y$ if $x - y$ is divisible by n. The set $\mathbb{R}$ of real numbers, which will be described in detail in Section 5, comes equipped with a tight apartness defined by setting $x \neq y$ if $|x - y| > 0$. The set $\mathbb{R}$ is not discrete unless LPO holds, as we noted in Section 1. Markov's principle is equivalent to the assertion that the inequality on $\mathbb{R}$ is the denial inequality.

A set is **nonempty**, or **nonvoid**, if we can construct an element of it; an **empty set** ϕ is a set that cannot be nonempty. To show that a set S is nonempty, it is not enough to prove that S cannot be empty: the set S of digits that appear infinitely often in the decimal expansion of π cannot be empty, at least from a classical point of view; but to show that S is nonempty, we must specify a digit that appears infinitely often.

If S_1 and S_2 are subsets of a set A such that $x \in S_2$ whenever $x \in S_1$, we say that S_1 is **contained**, or **included**, in S_2, and we write $S_1 \subset S_2$. Two subsets of A are **equal** if each is contained in the other. With this equality, the collection of all subsets of A is a set, called the **power set** $\mathcal{P}(A)$ of A.

A **function** from a set A to a set B is an algorithm f which produces an element $f(x)$ of B when applied to an element x of A, and which has the property that $f(x) = f(x')$ whenever $x = x'$; thus functions, like properties, are extensional. The notation $f:A \to B$ indicates that f is a function from A to B. A function is also called a **mapping**, or a **map**. The set A is called the **domain** of f, written **dom** f, and the set $\{f(x) : x \in A\}$, with equality inherited from B, is called the **range** of f, written **ran** f. If the range of f equals B, we say that f maps A **onto** B; if $x = x'$ whenever $f(x) = f(x')$, then f is **one-one**. The set B^A is the collection of all mappings from A to B, with $f = g$ if $f(x) = g(x)$ for all x in A.

A **partial function** f from a set A to a set B, or, by abuse of notation, a **partial function** $f:A \to B$, is a mapping from a subset of A into B. The **composition**, or **composite**, of partial functions $f:A \to B$ and $g:B \to C$ is the partial function $g \circ f:A \to C$ (sometimes written gf) defined by $g \circ f(x) \equiv g(f(x))$ wherever the right side exists.

A subset S of a set A is **detachable** from A if for each x in A either $x \in S$ or $x \notin S$. A detachable subset S of A may be identified with the function f from A to $\{0,1\}$ such that $f(x) = 1$ if and only if $x \in S$.

An (infinite) **sequence** x is a finite routine that associates an object x_n with each positive integer n; we often denote a sequence x by

$(x_n)_{n=1}^{\infty}$, or simply (x_n). If each term x_n belongs to a fixed set A, then the sequence (x_n) is a function from $\mathbb{N}^+$ to A. A **subsequence** of (x_n) consists of (x_n) and a sequence $(n_k)_{k=1}^{\infty}$ of positive integers such that $n_1 < n_2 < \ldots$; we identify this subsequence with the sequence whose k^{th} term is x_{n_k}.

Sometimes we speak of sequences defined on sets of the form $\{n \in \mathbb{Z} : n \geq N\}$, where N is fixed; such a sequence is written $(x_n)_{n=N}^{\infty}$ or, when the value of N is clearly understood, simply (x_n).

A **finite sequence of length n**, where n is a nonnegative integer, is a mapping x with domain $\{1,\ldots,n\}$; such a map can be identified with the **ordered n-tuple** $(x_1,\ldots,x_n)$, where $x_i \equiv x(i)$.

A set A is **finitely enumerable** if there exist a nonnegative integer n and a map f from $\{1,\ldots,n\}$ onto A. If the map f is also one-one, we say that A is **finite**, and that it **has n elements.** A finitely enumerable set is finite if and only if it is discrete under the denial inequality. Note that, according to our definitions, an empty set is both finite and finitely enumerable.

A set A is **countable** if there is a function from a detachable subset of $\mathbb{N}$ onto A. An empty set is countable, and a nonempty set is countable if and only if it is the range of a function with domain $\mathbb{N}$. We say that A is **countably infinite** if there is a one-one map from $\mathbb{N}$ onto A.

A **family of elements** of a set A is an **index set** I together with a function φ from I to X. We write φ_i for $\varphi(i)$, and denote the family by $(\varphi_i)_{i \in I}$. A family of elements of A with index set $\mathbb{N}^+$ is a sequence of elements of A.

Let $(S_i)_{i \in I}$ be a **family of subsets** of A – that is, a family of elements of $\mathcal{P}(A)$. The **union** of this family is the subset

$$\bigcup_{i \in I} S_i \quad = \quad \{x \in A : x \in S_i \text{ for some } i \text{ in } I\}$$

of A. The **intersection** of the family $(S_i)_{i \in I}$ is the subset

$$\bigcap_{i \in I} S_i \quad = \quad \{x \in A : x \in S_i \text{ for each } i \text{ in } I\}$$

of A. The **Cartesian product** $\Pi_{i \in I} S_i$ is the subset of A^I consisting of those functions f such that $f(i) \in S_i$ for each i in I. If I is the set of positive integers, or the set $\{1,2,\ldots,n\}$, then $(S_i)_{i \in I}$ is a sequence of sets, and the elements of $\Pi_{i \in I} S_i$ are also sequences; in this case, the S_i can be arbitrary sets, and need not be subsets of a previously defined set

A. If $I \equiv \{1,2,\ldots,n\}$, then we denote the product $\Pi_{i\in I} S_i$ by $S_1 \times S_2 \times\cdots\times S_n$, and refer to its elements $(x_1,x_2,\ldots,x_n)$ as **ordered** n**-tuples**, or, in the case $n = 2$, **ordered pairs.**

3. Informal intuitionistic logic

> *The belief in the universal validity of the principle of the excluded third in mathematics is considered by the intuitionists as a phenomenon of the history of civilization of the same kind as the former belief in the rationality of π, or in the rotation of the firmament about the earth.*
> (L.E.J. Brouwer, [**Brouwer**, page 7])

Constructive mathematics may be characterized by its *numerical content* and *computational method*. We have discussed numerical content in some detail; it is time we said a bit more about computational method.

The classical mathematician believes that every mathematical statement P is either true or false, whether or not he possesses a proof or disproof of P. Thus, in CLASS, Fermat's conjecture is either true or false, although we don't yet know which. The constructive mathematician does not consider a statement P to be true or false unless he can either prove it or disprove it. In this sense, Fermat's conjecture is considered neither true nor false, although it may be true tomorrow, or false tomorrow.

The constructive mathematician interprets the logical connectives and quantifiers according to **intuitionistic logic.** The important differences centre upon the connectives $\vee$ and $\Rightarrow$, and the quantifier $\exists$.

Suppose that in the middle of a computation we know that $P \vee Q$ is true. Then we should be able to continue our computation in one of two ways depending upon whether P is true or Q is true (we assume that if P and Q are both true, then we are content to proceed in either of the two possible ways). But this implies that if we assert $P \vee Q$, we must be prepared to assert P or to assert Q.

We may consider the statement $\exists x P(x)$ as a multiple disjunction, asserting that $P(x_1)$ or $P(x_2)$ or ... as x_i runs over the universe of discourse. If we assert $\exists x P(x)$, then we must be able to produce an element a such that $P(a)$. The assertion $P_1 \vee P_2$ may be interpreted as $\exists i\, P_i$, where the universe of discourse for i is $\{1,2\}$.

The statement $P \Rightarrow Q$ means that Q holds under the assumption that P holds: we show $P \Rightarrow Q$ by deriving Q from the hypothesis P. The statement $\neg P$ means that $P \Rightarrow Q$ where Q is contradictory (for example, Q is $(0 = 1)$). Note that the constructive interpretation of $P \Rightarrow Q$ is weaker than its classical interpretation, $\neg P \vee Q$, because from $\neg P \vee Q$ and P we can derive Q; but it is stronger than the intepretation $\neg(P \wedge \neg Q)$, because we can derive the contradiction $Q \wedge \neg Q$ from $P \Rightarrow Q$ and $P \wedge \neg Q$.

The classical principle $\neg\neg P \Rightarrow P$ is rejected in constructive mathematics because if $P \equiv \exists x Q(x)$, then $\neg\neg P$ merely says that it would be absurd to deny that there is x such that $Q(x)$, but gives no clue about how to construct x. On the other hand it is readily shown that

$$\neg\neg\neg P \Rightarrow \neg P.$$

For, under the assumption $\neg\neg\neg P$, we can prove $\neg P$ by deriving a contradiction from P: indeed, if P, then $\neg P$ is absurd; whence $\neg\neg P$, which contradicts $\neg\neg\neg P$. The principle $\neg\neg\neg P \Rightarrow \neg P$ was an early observation of Brouwer's.

Constructive mathematics is not based on a prior notion of logic; rather, our interpretations of the logical connectives and quantifiers grow out of our mathematical intuition and experience. Nevertheless, the formalization of a few basic axioms, and the subsequent axiomatic development of a logic for constructive mathematics – **first-order intuitionistic logic** - has had an impact on mathematical logic far exceeding any initial expectations. We shall return to this topic in Chapter 7.

4. Choice axioms

The full axiom of choice reads as follows.

If S is a subset of $A \times B$*, and for each* x *in A there exists* y *in B such that* $(x,y) \in S$*, then there is a function* f *from A to B such that* $(x,f(x)) \in S$ *for each* x *in A.*

The function f is called a **choice function** for S. Goodman and Myhill have shown that this axiom implies the law of excluded middle; so it cannot be a part of BISH. Their argument establishing $P \vee \neg P$ is very simple. Let $A \equiv \{s,t\}$, where $s = t$ if and only if P; let $B = \{0,1\}$; and let $S =$

$\{(s,0),(t,1)\} \subset A \times B$. If $f:A \to B$ is a choice function for S, then either

(i) $f(s) = 1$ or $f(t) = 0$, so that $s = t$, and therefore P holds; or else

(ii) $f(s) = 0$ and $f(t) = 1$, so that s cannot equal t, and therefore P cannot hold.

Although the full axiom of choice is not part of BISH, special cases are generally accepted and widely used by constructive mathematicians. One of these is the **axiom of countable choice**, which is the case $A \equiv \mathbb{N}$ of the full axiom of choice. Another is the stronger **axiom of dependent choice**, which reads as follows.

If $a \in A$ and $S \subset A \times A$, and for each x in A there exists y in A such that $(x,y) \in S$, then there exists a sequence of elements $a_1,a_2,\ldots$ of A such that $a_1 = a$ and $(a_n,a_{n+1}) \in S$ for each positive integer n.

A consequence of the intuitionistic interpretation of the term 'sequence' is that, in INT, the axiom of choice holds with $A \equiv \mathbb{N}^{\mathbb{N}}$, and hence with $A \equiv 2^{\mathbb{N}}$ (where, as is customary, 2 is identified with the set $\{0,1\}$). We shall discuss this further in Chapter 5.

5. Real numbers

A **rational number** is a pair of integers m/n, with $n \neq 0$. Two rational numbers m_1/n_1 and m_2/n_2 are **equal** if $m_1n_2 = m_2n_1$. The familiar operations and relations on the rational numbers remain unchanged in a constructive context.

Analysis begins with the real numbers. A **real number** is presented by rational approximations, and may be identified with a sequence $x = (x_n)$ of rational numbers that is **regular**, in the sense that

$$|x_m - x_n| \leq m^{-1} + n^{-1}$$

for all positive integers m and n. Any Cauchy sequence of rational numbers could be considered to be a real number, but by requiring that the sequence (x_n) be regular, we avoid having to provide an auxiliary sequence that tells how large n must be for x_n to approximate x to a specified accuracy.

Two real numbers x and y are **equal** if $|x_n - y_n| \leq 2/n$ for each positive integer n. Because x and y are regular, this is equivalent to demanding that $|x_n - y_n|$ be arbitrarily small for sufficiently large values of n. It then readily follows that equality is an equivalence relation; with this equality we have completed the construction of the set $\mathbb{R}$ of real numbers.

The **canonical bound** K_x of a regular sequence x of rational numbers is the least positive integer greater than $|x_1| + 2$. Clearly, $|x_n| < K_x$ for each n. The arithmetic operations on $\mathbb{R}$ are defined in terms of operations on regular sequences as follows:

(5.1)

$$\text{(i)} \quad (x \pm y)_n = x_{2n} \pm y_{2n}$$
$$\text{(ii)} \quad (xy)_n = x_{2kn}y_{2kn} \quad \text{where } k = \max\{K_x, K_y\}$$
$$\text{(iii)} \quad \max\{x,y\}_n = \max\{x_n, y_n\}$$
$$\text{(iv)} \quad \min\{x,y\}_n = \min\{x_n, y_n\},$$
$$\text{(v)} \quad |x|_n = |x_n|,$$

where, for example, $(x + y)_n$ denotes the nth term of the real number $x + y$. Note that $|x| = \max\{x,-x\}$. The set $\mathbb{Q}$ of rational numbers may be identified with a subset of $\mathbb{R}$ by associating with each rational number r the constant sequence $(r,r,r,\ldots)$.

A real number x is **positive** if there exists n such that $x_n > 1/n$. Thus the assertion that a given real number is positive is a simply existential statement. Because x is regular, it is straightforward to show that x is positive if and only if there exists a positive rational number ϵ such that $x_n > \epsilon$ for all sufficiently large values of n.

If x and y are real numbers, then we define $x < y$ to mean $y - x$ is positive. Thus $x > 0$ if and only if x is positive. The following properties of positive numbers are easily derived.

(5.2)

(i) *if* $x > 0$ *and* $y > 0$, *then* $x + y > 0$,

(ii) $\max\{x,y\} > 0$ *if and only if* $x > 0$ *or* $y > 0$,

(iii) *if* $x + y > 0$, *then* $x > 0$ *or* $y > 0$.

The **inequality** on $\mathbb{R}$ is given by

$$x \neq y \text{ if and only if } |x - y| > 0.$$

If $x \neq y$, then $x-y > 0$ or $y-x > 0$. Assuming, for example, that $x-y > 0$, for all real numbers z we have $(x-z) + (z-y) > 0$; whence $x-z > 0$ or $z-y > 0$, by (5.2,iii). Thus the inequality on $\mathbb{R}$ is an apartness.

A real number x is **negative** if $-x$ is positive, and is

nonnegative if $x_n > -1/n$ for each n. If x is nonnegative, then x_n is eventually larger than any given negative rational number, and conversely. As, for each n, either $x_n \leq 1/n$ or $-x_n > -1/n$, it follows that if x cannot be positive, then $-x$ is nonnegative.

If x and y are real numbers, then we define $x \leq y$ to mean that $y - x$ is nonnegative. Thus $x \geq 0$ if and only if x is nonnegative. Most of the algebraic and order properties of $<$ and $\leq$ carry over from classical to constructive mathematics. Among the more important properties that do not carry over are the following:

(a) $x > y$ or $x = y$ or $x < y$,

(b) $x \leq y$ or $y \leq x$,

(c) if $x \leq y$ is contradictory, then $x > y$.

Property (a) is equivalent to LPO, property (b) to LLPO, and property (c) to Markov's principle. The apartness property,

(5.3) if $x < y$, then $x < z$ or $z < y$,

of the inequality on $\mathbb{R}$ is commonly used as a constructive substitute for property (a).

It is easy to show that if $x \equiv (x_n)$ is a real number, then $|x - x_n| \leq n^{-1}$ for each n.

A real number x is **nonzero** if $x \neq 0$. For nonzero x, the **reciprocal** x^{-1}, or $1/x$, is defined as follows. Choose a positive integer m such that $|x_n| \geq 1/m$ for all $n \geq m$, and define $1/x$ by

$$\begin{aligned}(1/x)_n &= 1/x_{m^3} \quad \text{if } n < m, \\ &= 1/x_{nm^2} \quad \text{if } n \geq m.\end{aligned}$$

We leave it to the reader to show that x^{-1} is the unique real number t such that $xt = 1$; and that the map taking a nonzero real number x to x^{-1} maps onto the set of nonzero real numbers, and has the expected algebraic properties.

With the real numbers at hand, it is routine to construct the set $\mathbb{C}$ of complex numbers, together with its inequality.

PROBLEMS

1. Show that WLPO entails LLPO.

2. Show that the inequality on a discrete set is a tight apartness.

3. Let B be a set with an inequality, and let A be any set. For f and g in B^A, set $f \neq g$ if there exists x in A such that $f(x) \neq g(x)$. Show that this defines an inequality on B^A. Show that if the inequality on B is a (tight) apartness, then so is the inequality on B^A.

4. Show that the set of binary sequences, with the inequality of Problem 2, is discrete if and only if LPO holds.

5. Construct a Brouwerian example of a finitely enumerable set that is not finite.

6. Prove that the axiom of dependent choice implies the axiom of countable choice. Prove that the axiom of countable choice implies the axiom of dependent choice in the case $A \equiv \mathbb{N}$.

7. Let A be a subset of a set B. A subset A' of B is a (strong) **complement** of A if $A \cup A' = B$ and $A \cap A' = \phi$. Following Diaconescu, show that the axiom of choice implies that every subset has a complement, by letting $C \equiv B_1 \cup B_2$ be the disjoint union of two copies of B, and D be C with corresponding elements of A_1 and A_2 identified.

8. Show that the following are equivalent in BISH.
   (i) Markov's principle.
   (ii) For each real number x, if $\neg(x = 0)$, then $x \neq 0$.
   (iii) For each real number x, if $\neg(x = 0)$, then there exists a real number y such that $xy = 1$.

9. Show that if every one-one continuous map $f:[0,1] \to \mathbb{R}$ has the property that $f(x) \neq f(y)$ whenever $x \neq y$, then Markov's principle holds.

10. Give Brouwerian counterexamples to each of the following statements.
    (i) Every real number is either rational or irrational.
    (ii) If f is a uniformly continuous map of $[0,1]$ into $\mathbb{R}$ such that $f(0)f(1) < 0$, then $f(x) = 0$ for some x with $0 < x < 1$.
    (iii) Every ideal in the ring of integers is finitely generated.

11. Construct a Brouwerian example of a real number that does not have a binary expansion.

12. For real numbers x and y, show that if $xy \neq 0$, then $x \neq 0$ and $y \neq 0$. Give a Brouwerian counterexample to the statement that if $xy = 0$, then $x = 0$ or $y = 0$.

13. Show that a nonempty set is countable if and only if it is the range of a mapping from the positive integers.

14. A function $f : \mathbb{R} \to \mathbb{R}$ is said to **jump on the right of** a if there exists $\epsilon > 0$ such that $|f(x) - f(a)| \geq \epsilon$ for all $x > a$. Show that the existence of such a function entails LPO.

15. Give a Brouwerian counterexample to the statement that if r_1, r_2, and r_3 are real roots of a monic quadratic polynomial with coefficients in $\mathbb{R}$, then $r_i = r_j$ for some $i \neq j$.

16. Show that the intersection of two countable subsets of $\mathbb{N}$ is countable.

NOTES

For Bishop's views on mathematics see the Preface and Chapter 1 of **Bishop**, and the articles *Schizophrenia in contemporary mathematics* (American Math. Soc. Colloquium Lectures 1973, Univ. of Montana, Missoula) and *The crisis in contemporary mathematics* (Historia Math. 2 (1975), 507–517). Good references for intuitionism are **Dummett** and **Brouwer**. References on authentic Russian constructivism make difficult reading; perhaps the most approachable is **Kushner**; a more readable, but less orthodox, alternative is Aberth's *Computable Analysis* (McGraw-Hill, 1980).

The restriction that the passage from one step to another in a finite routine be deterministic is relaxed in some presentations of intuitionistic mathematics – see Chapter 5.

Brouwer's notion of a **fleeing property** corresponds to a binary sequence a for which it is not known whether $a_n = 1$ for some n, or whether $a_n = 0$ for all n. Brouwer called a fleeing property **two-sided with regard to parity** if the conclusion of LLPO had not been demonstrated for it (see **Brouwer**, page 6).

Heyting gives the example of the set of digits that appear infinitely often in the decimal expansion of π in *Intuitionistic views on the nature of mathematics*, Synthese 27 (1974), 79–91.

Bishop calls *subfinite* the sets that we call *finitely enumerable*. Note that subsets of finite sets need not be finitely enumerable.

Bishop calls *surjective* the maps that we call *onto*, reserving the term *onto* for a map that admits a right inverse (a cross-section).

Several authors, including Myhill (*Constructive set theory*, J. Symbolic Logic 40 (1975), 347–382), have objected to the use of the power set construction within constructive mathematics. See also pages 19 and 191 of **Beeson**.

Brouwer uses the term 'inhabited' where we use 'nonempty'. In **Bishop** and **Bishop-Bridges**, a countable set is nonempty. With the definition given here, which agrees with that in **Brouwer** for discrete sets, we need not be able to decide whether a given countable set is nonempty.

Note that our definition permits a finite sequence to be empty — that is, have no terms.

The interpretation of $\Rightarrow$ has always been controversial, and this remains the case in the constructive context. The intuitionists interpret $P \Rightarrow Q$ proof-theoretically, to mean that we have a finite routine for converting any proof of P to a proof of Q. For a discussion of another interpretation, due to Gödel, see Bishop's paper *Mathematics as a numerical language* (in *Intuitionism and Proof Theory*, North-Holland, 1971); for a deeper analysis of the intuitionistic interpretations of the logical connectives and quantifiers, see **Dummett**.

Brouwer referred to the principle $\neg\neg P \Rightarrow P$ as *the simple principle of the reciprocity of absurdity* (**Brouwer**, page 11), because it says, essentially, that if Q is equivalent to the absurdity of P, then P is equivalent to the absurdity of Q. Brouwer disliked logical shorthand and expressed the theorem $\neg\neg\neg P \Rightarrow \neg P$ by *absurdity of absurdity of absurdity is equivalent to absurdity* (**Brouwer**, page 12).

Although we refer to 'choice axioms', the practitioners of constructive mathematics do not see themselves as exercising choice; rather, they see the choice function as inherent in the *meaning* of the statement $\forall n \exists m P(n,m)$.

In *Axiom of choice and complementation* (Proc. Amer. Math. Soc. 51 (1975), 176–178), Diaconescu shows that in a topos-theoretic setting, the axiom of choice implies the law of excluded middle. The set-theoretic version that we use is found in *Choice implies excluded middle*, by Goodman and Myhill (Z. Math. Logik Grundlagen Math. 23 (1978), 461).

Chapter 2. Constructive Analysis

In which various examples are given to illustrate techniques and results in constructive analysis. Particular attention is paid to results which are classically trivial, or have little or no classical content, but whose constructive proof requires considerable ingenuity. The final section draws together various ideas from earlier parts of the chapter, and deals with a result which is interesting both in itself and for the questions raised by its proof.

1. Complete metric spaces

The appropriate setting for constructive analysis is in the context of a metric space: there is, as yet, no useful constructive notion corresponding to a general topological space in the classical sense.

We assume that the reader is familiar with the elementary classical theory of metric spaces and normed linear spaces. Most definitions carry over to the constructive setting, although we may have to be careful which of several classically equivalent definitions we use. For example, some classical authors define a closed set to be one whose complement is open, whereas others define a closed set S to be one containing all limits of sequences in S; it turns out that the latter is the more useful notion, and hence the one adopted as definitive, in the constructive setting.

We shall use $B(a,r)$ (respectively, $\overline{B}(a,r)$) to denote the open (respectively, closed) ball with centre a and radius r in a metric space.

The prototype complete space is the real line $\mathbb{R}$.

(1.1) Theorem *$\mathbb{R}$ is a complete metric space.*

Proof. Consider any Cauchy sequence (x_n) in $\mathbb{R}$. For each positive integer k, choose N_k in $\mathbb{N}^+$ such that

$$|x_m - x_n| \leq k^{-1} \qquad (m,n \geq N_k).$$

Write

$$\upsilon(k) \equiv \max\{3k, N_{2k}\} \qquad (k \in \mathbb{N}^+).$$

Let x_k^∞ be the $2k^{\text{th}}$ term of the regular sequence $x_{\upsilon(k)}$. Then we have

$$\begin{aligned}|x_m^\infty - x_n^\infty| &\leq |x_m^\infty - x_{\upsilon(m)}| + |x_{\upsilon(m)} - x_{\upsilon(n)}| + |x_{\upsilon(n)} - x_n^\infty| \\ &\leq (2m)^{-1} + (2m)^{-1} + (2n)^{-1} + (2n)^{-1} = m^{-1} + n^{-1}.\end{aligned}$$

Hence $x^\infty \equiv (x_n^\infty)_{n=1}^\infty$ is a real number. Moreover, if $n \geq \upsilon(k)$, then

$$\begin{aligned}|x^\infty - x_n| &\leq |x^\infty - x_n^\infty| + |x_n^\infty - x_{\upsilon(n)}| + |x_{\upsilon(n)} - x_n| \\ &\leq n^{-1} + (2n)^{-1} + (2k)^{-1} \\ &\leq (3k)^{-1} + (6k)^{-1} + (2k)^{-1} = k^{-1}.\end{aligned}$$

Thus (x_n) converges to the real number x^∞. □

For those concerned with countable choice, there are two instances of that axiom in the above proof. The first occurs when we construct the sequence (N_k); the second when we construct the sequence (x_k^∞) by treating the sequence (x_n) of real numbers as if it were a sequence of regular sequences of rational numbers.

A real number b is called the **supremum**, or **least upper bound**, of a subset S of $\mathbb{R}$ if it is an upper bound of S, and if, for each $\epsilon > 0$, there exists x in S such that $x > b - \epsilon$. It is readily shown that the supremum of S, if it exists, is unique; we denote it by **sup S**. The definition of **infimum**, or **greatest lower bound**, of S (written **inf S**) is left to the reader.

It is trivial to show that LPO is a consequence of the classical **least-upper-bound principle**: a nonvoid set of real numbers that is bounded above has a supremum. As an application of the completeness of $\mathbb{R}$ we have the following **constructive least-upper-bound principle**.

(1.2) Theorem *Let* S *be a nonvoid set of real numbers that is bounded above. Then* sup S *exists if and only if for all* α,β *in* $\mathbb{R}$ *with* $\alpha < \beta$, *either* β *is an upper bound of* S *or there exists* x *in* S *with* $x > \alpha$.

Proof. If $M \equiv \sup S$ exists and $\alpha < \beta$, then either $M < \beta$ or $M > \alpha$; in the latter case, we can find x in S with $x + (M - \alpha) > M$ and therefore $x > \alpha$.

Assume, conversely, that the stated condition holds. Choose an upper bound u_0 of S, and $\epsilon > 0$, so that $[u_0-\epsilon, u_0] \cap S$ is nonvoid. With

$\epsilon_n \equiv \epsilon(3/4)^n$, construct a sequence (u_n) of upper bounds of S inductively, such that for each n,

(i) $[u_n-\epsilon_n, u_n] \cap S$ is nonvoid;

(ii) either $u_{n+1} = u_n - \epsilon_n/4$ or $u_{n+1} = u_n$.

To do so, assume that $u_0,\ldots,u_n$ have been constructed with the appropriate properties. Then either $u_n - \epsilon_n/4$ is an upper bound of S, or else there exists ξ in S with $\xi > u_n - \epsilon_{n+1}$. In the first case, set $u_{n+1} \equiv u_n - \epsilon_n/4$; in the second, set $u_{n+1} \equiv u_n$. This completes the inductive construction.

Using (ii) we easily see that

$$|u_m - u_n| \leq \Sigma_{i=n}^{\infty}\epsilon_i/4 = \epsilon_n$$

whenever $m \geq n$. Then (u_n) is a Cauchy sequence, so its limit u_∞ exists by (1.1). Clearly $u_n - \epsilon_n \leq u_\infty \leq u_n$ for each n. Since each u_n is an upper bound of S, so is u_∞. Therefore, by (i), $[u_n-\epsilon_n, u_\infty] \cap S$ is nonvoid for each n; so u_∞ is the supremum of S. □

This last proof is an illustration of dependent choice: the element u_{n+1} is not unambiguously specified, as the two conditions in question may both hold and the possible choices for u_{n+1} depend on u_n. However we can modify the proof so that only countable choice is involved (Problem 1).

A fundamental property of complete metric spaces is given by **Baire's theorem:**

(1.3) Theorem *The intersection of a sequence of dense open sets in a complete metric space X is dense in X.*

The standard classical proof of this theorem carries over into the constructive setting. We shall return to that proof, which involves dependent choice, in Section 2 below.

Let X be a set with an inequality $\neq$. If for each sequence (x_n) of elements of X there exists x in X such that $x \neq x_n$ for each n, then X is said to be **uncountable**. Using Baire's theorem, we can prove **Cantor's theorem.**

(1.4) Theorem. *$\mathbb{R}$ is uncountable; in fact, if (x_n) is any sequence of real numbers, then $\{x \in \mathbb{R}: x \neq x_n$ for all $n\}$ is dense in $\mathbb{R}$.*

Proof. Consider any sequence (x_n) of real numbers, and for each n in $\mathbb{N}$ define

$$U_n \equiv \{x \in \mathbb{R}: x \neq x_n\}.$$

Then $B(x,|x - x_n|) \subset U_n$ for all x in U_n, so that U_n is open. To show that U_n is dense in $\mathbb{R}$, consider any real number x. For each $\epsilon > 0$, either $x_n > x$, in which case $x \in U_n$; or $x_n < x + \epsilon$, so that $x + \epsilon \in U_n$ and $|x - (x + \epsilon)| = \epsilon$. Thus there are elements of U_n arbitrarily close to x, and so U_n is dense in $\mathbb{R}$. Applying Baire's theorem, we see that

$$\{x \in \mathbb{R} : x \neq x_n \text{ for all } n\} = \bigcap_{n=1}^{\infty} U_n$$

is dense in $\mathbb{R}$. □

2. Baire's theorem revisited

We often find that a single classical theorem gives rise to several constructively inequivalent results. A particularly common example of this occurs when a classical theorem $p \Rightarrow q$ is written as $q' \Rightarrow p'$, where p',q' are classically, but not constructively, equivalent to $\neg p,\neg q$ respectively. To illustrate this, we prove versions of Baire's theorem that are classically equivalent to the contrapositive of (1.3), and therefore to (1.3) itself, but whose constructive content is altogether different from that of (1.3). Before doing so, however, we must introduce several notions, the first of which has no classical significance but is often essential in constructive analysis.

A subset S of a metric space (X,ρ) is **located** in X if the **distance**

$$\rho(x,S) \equiv \inf\{\rho(x,s) : s \in S\}$$

from x to S exists for each x in X. Note that if S is located, then so is the closure $\overline{S}$ of S in X; if also T is dense in S, then T is located in X, and $\rho(x,T) = \rho(x,S)$ for each x in X. There are Brouwerian examples of subsets of $\mathbb{R}$ that are not located (Problem 3).

The **metric complement** of a set S in a metric space X is the set $X - S$ of all points x of X that are **bounded away** from S, in the sense that $\rho(x,s) \geq \alpha$ for some $\alpha > 0$ and all s in S. Note that $X - S$ is open in X; and that if S is located in X, then $X - S = \{x \in X : \rho(x,S) > 0\}$.

By a construction similar to that extending $\mathbb{Q}$ to $\mathbb{R}$, an arbitrary metric space X can be embedded isometrically as a dense subset of a complete metric space $\hat{X}$, called the **completion** of X; then X is a complete metric space if and only if $X = \hat{X}$.

The essence of the standard proof of Baire's theorem is the following lemma.

(2.1) Lemma *Let (U_n) be a sequence of dense open sets in a metric space X, $U \equiv \bigcap_{n=1}^{\infty} U_n$, x_0 a point of X, and $r_0 > 0$. Then there exists a point x_∞ of the completion of X such that* (i) $\rho(x_\infty, x_0) \leq r_0$, and (ii) *if $x_\infty \in X$, then $x_\infty \in U$.*

Proof. Since U_1 is dense and open in X, there exist x_1 in X and a real number r_1 such that $0 < r_1 < 2^{-1}$ and $\overline{B}(x_1, r_1) \subset U_1 \cap B(x_0, r_0)$. Similarly, there exist x_2 in X and a real number r_2 such that $0 < r_2 < 2^{-2}$ and $\overline{B}(x_2, r_2) \subset U_2 \cap B(x_1, r_1)$. Carrying on in this way, we construct inductively a sequence $(x_n)_{n=0}^{\infty}$ in X, and a sequence $(r_n)_{n=0}^{\infty}$ of positive numbers converging to 0, such that if $m \geq n \geq 1$, then $x_m \in \overline{B}(x_n, r_n) \subset U_n \cap B(x_{n-1}, r_{n-1})$. Let x_∞ be the limit of (x_n) in the completion $\hat{X}$ of X. Then x_∞ belongs to the closure of $B(x_0, r_0)$ in $\hat{X}$, so that $\rho(x_\infty, x_0) \leq r_0$. Also, if $x_\infty \in X$, then for each $n \geq 1$, x_∞ belongs to the closed subset $\overline{B}(x_n, r_n)$ of X, and therefore to U_n; whence $x_\infty \in U$. □

We can now give the

Proof of Baire's theorem (1.3). Given a sequence (U_n) of dense open subsets of a complete metric space X, let x_0 be any point of X, and r_0 any positive number. Construct the point x_∞ of the completion of X as in Lemma (2.1). Since X is complete, x_∞ belongs to X and therefore to $U \cap \overline{B}(x_0, r_0)$, where $U \equiv \bigcap_{n=1}^{\infty} U_n$. As x_0 and r_0 are arbitrary, it follows that U is dense in X. □

A metric space X is said to be **incomplete** if there exists a Cauchy sequence (x_n) in X that is **eventually bounded away** from any given point of X, in the sense that for each x in X there exist N in $\mathbb{N}$ and $\alpha > 0$ such that $\rho(x, x_n) \geq \alpha$ for all $n \geq N$. Note that X is incomplete if and only if there exists a point ξ in the completion of X such that $\rho(\xi, x) > 0$ for each x in X.

Since Markov's principle is true in CLASS and RUSS, the next theorem can be viewed as another constructive version of Baire's theorem.

(2.2) Theorem *Markov's principle is equivalent to the following statement:*

(2.2.1) *If (U_n) is a sequence of dense open sets in a metric space X, such that the metric complement of $\bigcap_{n=1}^{\infty} U_n$ is nonvoid, then X is incomplete.*

Proof. Assume that Markov's principle holds, and let (U_n) be a sequence of dense open subsets of a metric space X such that the metric complement of $U \equiv \bigcap_{n=1}^{\infty} U_n$ is nonvoid. Choose x_0 in $X{-}U$ and $r_0 > 0$ such that $\rho(x,x_0) \geq 2r_0$ for all x in U, and construct the point x_∞ in the completion of X as in (2.1). If $x \in X$ and $x_\infty = x$, then by (2.1), $x_\infty \in U$ and $\rho(x_\infty,x_0) \leq r_0$, which contradicts our choice of r_0. Thus for all x in X, $\neg(x_\infty = x)$, and therefore $\rho(x_\infty,x) > 0$ by Markov's principle. Hence X is incomplete.

Conversely, assume (2.2.1), and consider an increasing binary sequence (a_n) such that $\neg\forall n(a_n = 0)$. Let $0 = r_0, r_1, \ldots$ be an enumeration of the rational points of $[0,1]$, and consider the metric space

$$X \equiv \{0\} \cup \{x \in \mathbb{Q} : \exists n(a_n = 1 \text{ and } 0 \leq x \leq 1/n)\},$$

with the metric inherited from $\mathbb{R}$. For each n, define

$$U_n \equiv X - \{r_{n-m} : m \leq n,\ a_m = 1\}.$$

Then U_n is dense and open in X: for if $a_n = 0$, then $U_n = X$; while if $a_n = 1$, then $X = \mathbb{Q} \cap [0,1/N]$ for some N in $\mathbb{N}^+$, and U_n is the metric complement of a finite set in X. If there exists n with $a_n = 1$, then $U \equiv \bigcap_{n=1}^{\infty} U_n$ is empty; so if U is nonvoid, then $a_n = 0$ for all n, a contradiction. Thus U is empty, whence 0 is in its metric complement. By (2.2.1), X is incomplete, and so there exists a Cauchy sequence in X that is eventually bounded away from 0. Hence X contains a nonzero element, and therefore $a_n = 1$ for some n. □

The way in which the desired conclusion is reached in the second half of the proof of Theorem (2.2) is perhaps surprising. A more dramatic instance of this type of argument will be used to establish Theorem (6.1) below.

Theorem (2.2) does not seem to have any interesting applications in constructive mathematics. The following lemma will enable us to prove two more versions of Baire's theorem, each quite different from Theorems (1.3) and (2.2); the second of these new versions has at least one interesting application.

(2.3) Lemma *Let (U_n) be a sequence of located open sets in a complete metric space X, such that the metric complement of $U \equiv \bigcap_{n=1}^{\infty} U_n$ is nonvoid. Then there exist a point x_∞ in $X - U$ and an increasing binary sequence (λ_n) such that for each n,*

(i) *if $\lambda_n = 0$, then $x_\infty \in U_n$,*

(ii) *if $\lambda_n = 1$, then $X - U_k$ is nonvoid for some $k \leq n$.*

Proof. Choose x_0 in X and $r_0 > 0$ such that $\rho(x_0,x) \geq r_0$ for all x in U. Construct an increasing binary sequence $(\lambda_n)_{n=0}^{\infty}$ with $\lambda_0 = 0$, a sequence $(x_n)_{n=0}^{\infty}$ in X, and a sequence $(r_n)_{n=0}^{\infty}$ of real numbers, such that for each $n \geq 1$,

(a) $0 < r_n \leq r_{n-1}/3$,

(b) $\lambda_n = 0 \Rightarrow \rho(x_n,x_{n-1}) < r_{n-1}/3$ and $B(x_n,r_n) \subset U_n$,

(c) $\lambda_n = 1 \Rightarrow x_n = x_{n-1}$ and $X - U_k$ is nonvoid for some $k \leq n$.

We proceed by induction. Assume that we have constructed λ_n, x_n, and r_n. If $\lambda_n = 1$, set

(2.3.1) $$\lambda_{n+1} \equiv 1, \quad x_{n+1} \equiv x_n, \quad \text{and} \quad r_{n+1} \equiv r_n/3.$$

Otherwise, $\lambda_n = 0$ and either $\rho(x_n,U_{n+1}) > 0$ or $\rho(x_n,U_{n+1}) < r_n/3$. In the first case, define λ_{n+1}, x_{n+1}, and r_{n+1} by (2.3.1). In the second case, set $\lambda_{n+1} \equiv 0$, and choose x_{n+1} in U_{n+1} and r_{n+1} with $0 < r_{n+1} < r_n/3$, such that $\rho(x_n,x_{n+1}) < r_n/3$ and $B(x_{n+1},r_{n+1}) \subset U_{n+1}$. This completes the induction.

If $m > n \geq 0$, we now have

$$\rho(x_m,x_n) \leq \sum_{i=n+1}^{m} \rho(x_{i-1},x_i) < \sum_{i=n+1}^{m} r_{i-1}/3$$
$$\leq (r_n/3)\sum_{i=0}^{\infty}(1/3)^i = r_n/2.$$

Hence (x_n) is a Cauchy sequence in X, and so converges to a limit x_∞ in X such that $\rho(x_\infty,x_n) \leq r_n/2$ for all $n \geq 0$. Property (i) follows from this and (b); property (ii) follows from (c). Moreover, as $\rho(x_\infty,x_0) \leq r_0/2$, we have $\rho(x_\infty,x) \geq r_0/2$ for all x in U; whence $x_\infty \in X - U$. □

(2.4) Theorem *Markov's principle is equivalent to the following statement:*

(2.4.1) *If (U_n) is a sequence of located open sets in a complete metric space X, such that the metric complement of $U \equiv \bigcap_{n=1}^{\infty} U_n$*

is nonvoid, then $X - U_n$ *is nonvoid for some* n.

Proof. Assume Markov's principle, and let (U_n) satisfy the hypotheses of (2.4.1). Construct x_∞ and (λ_n) as in (2.3). Then if $\lambda_n = 0$ for all n, we have $x_\infty \in U$, a contradiction. Hence $\lambda_N = 1$ for some N, and therefore $X - U_n$ is nonvoid for some $n \leq N$.

Conversely, assume (2.4.1), and let (x_n) be any sequence in the complete metric space $X \equiv \{0,1\}$ such that $\neg\forall n(x_n = 0)$. It is easily shown that

$$U_n \equiv \{x \in X : x \geq x_n\}$$

is located and open in X for each n, that $1 \in U \equiv \bigcap_{n=1}^{\infty} U_n$, and that if $0 \in U$, then $\forall n(x_n = 0)$. Hence $\rho(0,x) = 1$ for all x in U, and so $0 \in X - U$. By (2.4.1), there exists n such that $X - U_n$ is nonvoid; whence $x_n = 1$. Thus Markov's principle holds. □

The proof of Theorem (2.4) illustrates a technique widely employed in the presence of completeness. Given a complete metric space X and properties $P(n)$, $Q(n)$ of a positive integer n, such that for each n either $P(n)$ or $Q(n)$ holds, we want to show that $Q(n)$ holds for some n. We construct an increasing binary sequence (λ_n) such that $P(n)$ holds if $\lambda_n = 0$, and $Q(n)$ holds if $\lambda_n = 1$. Using the information provided by the properties $P(n)$ and $Q(n)$, we then construct a Cauchy sequence (x_n) in X such that $x_n = x_{n-1}$ if $\lambda_n = 1$, and such that the behaviour of the limit of (x_n) in X will enable us to compute a value of n for which $\lambda_n = 1$ and therefore $Q(n)$ holds.

We shall return to this technique in Section 3. In the meantime, we have two more definitions.

A subset S of a metric space X is **colocated** in X if its metric complement $X - S$ is located; and S is **bilocated** in X if it is both located and colocated in X.

Lemma (2.3) also leads to a version of Baire's theorem that does not involve Markov's principle.

(2.5) Theorem *Let* (C_n) *be a sequence of closed, bilocated subsets of a complete metric space* X*, such that* $X = \bigcup_{n=1}^{\infty} C_n$. *Then* C_n *has nonvoid interior for some* n.

Proof. The proof is left as an exercise (Problem 5). □

We invite the reader to examine the standard classical proof of the open mapping theorem for bounded linear maps, in order to confirm that, in the light of Theorem (2.5), we can establish the following constructive version of the **open mapping theorem.**

(2.6) Theorem *Let u be a bounded linear mapping of a Banach space X onto a Banach space Y, such that $u(B(0,1))$ is bilocated in Y. Then u is an open mapping – that is, u maps open subsets of X onto open subsets of Y.* □

3. Located subsets

As Theorems (2.4) and (2.5) suggest, locatedness is a significant property of a subset of a metric space, and one which merits further investigation.

Even if a subset is located, we may not be able to prove that its metric complement is located (Problem 8). In certain cases, however, we can do so.

(3.1) Proposition. *If S is a located, convex subset of $\mathbb{R}^n$ such that $\mathbb{R}^n-S$ is nonvoid, then $\mathbb{R}^n-S$ is located.*

Proof. As the reader may verify, it is enough to prove that $T \equiv \mathbb{R}^n-S$ is located relative to the metric ρ_1 on $\mathbb{R}^n$, where

$$\rho_1(x,y) \equiv \sup\{|x_i - y_i| : 1 \leq i \leq n\}$$

for all $x \equiv (x_1,\ldots,x_n)$ and $y \equiv (y_1,\ldots,y_n)$ in $\mathbb{R}^n$. Accordingly, consider any x in $\mathbb{R}^n$, and for each $r > 0$ let $\overline{B}_1(x,r)$ be the closed ball with centre x and radius r relative to the metric ρ_1. Then $\overline{B}_1(x,r)$ is an n-dimensional cube with centre x and sides of length $2r$. Consider any two real numbers α,β such that $\alpha < \beta$. Let $v(1),\ldots,v(2^n)$ be the vertices of $\overline{B}_1(x,\frac{1}{2}(\alpha+\beta))$. It is left as an exercise for the reader to show that for all sufficiently small $\delta > 0$, and all points $w(1),\ldots,w(2^n)$ such that $\rho_1(v(i),w(i)) < \delta$ for all i in $\{1,\ldots,2^n\}$, the convex hull of $\{w(1),\ldots,w(2^n)\}$ contains $\overline{B}_1(x,\alpha)$. Fix such a number $\delta < \frac{1}{2}(\beta-\alpha)$. Either $\rho_1(v(i),S) < \delta$ for all i, or $\rho_1(v(i),S) > 0$ for some i. In the first case, for each i in $\{1,\ldots,2^n\}$ choose $w(i)$ in S so that $\rho_1(v(i),w(i)) < \delta$; then S contains the convex hull of $\{w(1),\ldots,w(2^n)\}$ and therefore contains $\overline{B}_1(x,\alpha)$. Thus $\rho_1(x,y) \geq \alpha$ for all y in T. On the other hand, if, for some i, we have $\rho_1(v(i),S) > 0$, then $v(i) \in T$ and

$\rho_1(x,v(i)) \le \frac{1}{2}(\alpha + \beta) < \beta$. Thus $\rho_1(x,S)$ exists, by the least-upper-bound principle (1.2). □

Two subsets S and T of $\mathbb{R}^n$ are said to **interlace** if there exist s,s' in S, $\lambda > 1$, and μ in (0,1) such that both $\lambda s + (1 - \lambda)s'$ and $\mu s + (1 - \mu)s'$ belong to T. If S and T are disjoint, and either S or T is convex, then S and T do not interlace. Thus Proposition (3.1) is a special case of the following general result, which is proved in the paper *Locating metric complements in* $\mathbb{R}^n$, in **Springer 873.**

(3.2) Theorem *If S is a located subset of* $\mathbb{R}^n$ *with nonvoid metric complement T, such that S and T do not interlace, then T is located.* □

The completeness technique used in the proof of Theorem (2.5) can be applied to produce a very useful result with almost no classical content.

(3.3) Lemma *Let S be a complete, located subset of a metric space X, and x a point of X. Then there exists y in S such that* $\rho(x,y) > 0$ *entails* $\rho(x,S) > 0$.

Proof. Construct a binary sequence (λ_n) such that

$$\lambda_n = 0 \Rightarrow \rho(x,S) < n^{-1},$$

and

$$\lambda_n = 1 \Rightarrow \rho(x,S) > (n+1)^{-1}.$$

Note that (λ_n) is increasing. Let z be any point of S. If $\lambda_1 = 1$, set $x_1 \equiv z$; if $\lambda_n = 0$, choose x_n in S with $\rho(x,x_n) < n^{-1}$; if $n > 1$ and $\lambda_n = 1$, set $x_n \equiv x_{n-1}$. Then (x_n) is a Cauchy sequence in X – in fact,

$$\rho(x_m,x_n) < 2n^{-1} \qquad (m \ge n).$$

As X is complete, (x_n) converges to a limit y in X satisfying $\rho(y,x_n) \le n^{-1}$ for all n. If $\rho(x,y) > 0$, choose N in $\mathbb{N}^+$ so that $\rho(x,y) > 3N^{-1}$. Then

$$\rho(x,x_N) \ge \rho(x,y) - \rho(y,x_N) > N^{-1},$$

so that $\lambda_N \ne 0$, and therefore $\lambda_N = 1$. Hence $\rho(x,S) > (N+1)^{-1} > 0$. □

To prove (3.3) classically, we choose y to be x if $\rho(x,S) = 0$; otherwise, we choose y to be any point in S.

(3.4) Corollary *If S is a complete, located subset of a metric space X and x is a point of X such that $\rho(x,y) > 0$ for all y in S, then* $\rho(x,S) > 0$. □

Corollary (3.4) provides conditions under which a uniforml continuous, positive-valued function – namely, the function taking y t $\rho(x,y)$ on S – has positive infimum. As we shall see in Chapter 6, a arbitrary uniformly continuous mapping of a compact metric space into ℝ need not have positive infimum.

4. Totally Bounded Spaces

Many of the most important results of classical analysi depend on an application of the property of compactness of a metric spac X. This property is usually defined in one of two classically equivaler ways. The first – every open cover of X contains a finite subcover – i of no use in constructive analysis as, for reasons that will be clea after Chapters 3 and 5, we cannot prove that the interval [0,1] is compac in that sense. The second notion, that of sequential compactness (ever sequence of points of X contains a convergent subsequence), i inapplicable because even the two-element set {0,1} cannot be shown to b sequentially compact.

There is, however, a third property which, for a metric space is classically equivalent to compactness in either of the above senses and which is widely applicable in constructive mathematics. It is th third notion that we shall dignify by the title of 'compactness'.

Let (X,ρ) be a metric space. An **ϵ-approximation** to X is subset Y of X such that for each x in X there exists y in Y wi $\rho(x,y) < \epsilon$. The space X is said to be **totally bounded** if for each $\epsilon >$ there exists a finite ϵ-approximation to X. A complete, totally bound metric space is said to be **compact**.

A totally bounded space X is **bounded**, in the sense th $\rho(x,y) \leq c$ for some c in ℝ and all x,y in X. A dense subset of a total bounded space is totally bounded.

A bounded, closed interval is compact relative to the standa metric on ℝ. More generally, closed balls in $\mathbb{R}^n$ and $\mathbb{C}^n$ are total bounded.

Recall that a set X is **finite**, under the denial inequality, if there is a one-one mapping φ of $\{1,\ldots,n\}$ onto X for some nonnegative integer n. A finite metric space is said to be **metrically finite** if the distance between any two distinct points is positive; that is, if the denial inequality is the same as the inequality induced by the metric. In the definition of a totally bounded space, the finiteness of the ϵ-approximations can be either relaxed to finite enumerability, or strengthened to metric finiteness:

(4.1) Lemma *Let ϵ be a positive number, and X a metric space with a finitely enumerable ϵ-approximation. Then X has a metrically finite η-approximation for each $\eta > \epsilon$.*

Proof. Let $\{x_1,\ldots,x_n\}$ be an ϵ-approximation to X, and let $\eta > \epsilon$. We proceed by induction on n. Either $\rho(x_i,x_j) > 0$ whenever $i \neq j$, or $\rho(x_i,x_j) < \eta - \epsilon$ for some pair of distinct indices i,j. In the former case, $\{x_1,\ldots,x_n\}$ is a metrically finite η-approximation. In the latter case, we can delete x_j from $\{x_1,\ldots,x_n\}$ and get a finitely enumerable ϵ'-approximation with $\epsilon' = \epsilon + \rho(x_i,x_j) < \eta$. □

A metric space is **separable** if it contains a countable dense set.

(4.2) Lemma *A totally bounded space is separable.*

Proof. If, for each positive integer n, F_n is a finite n^{-1}-approximation to the metric space X, then $F \equiv \bigcup_{n=1}^{\infty} F_n$ is a countable dense subset of X. □

Many applications of total boundedness depend on the next result.

(4.3) Proposition *If X is a totally bounded space, and f a uniformly continuous map of X into a metric space X', then $f(X) \equiv \{f(x) : x \in X\}$ is totally bounded.*

Proof. Consider any $\epsilon > 0$, and choose $\delta > 0$ such that $\rho(f(x),f(y)) < \epsilon$ whenever $x,y \in X$ and $\rho(x,y) < \delta$. Construct a δ-approximation $\{x_1,\ldots,x_n\}$ to X. Then for each x in X there exists i such that $\rho(x,x_i) < \delta$ and therefore $\rho(f(x),f(x_i)) < \epsilon$. Thus $\{f(x_1),\ldots,f(x_n)\}$ is a finitely enumerable ϵ-approximation to $f(X)$. Reference to (4.1) completes the proof. □

To apply Proposition (4.3), we need a lemma.

(4.4) Lemma *If S is a nonvoid totally bounded subset of* $\mathbb{R}$, *then* inf S *and* sup S *exist.*

Proof. Let α,β be real numbers with $\alpha < \beta$, and set $\epsilon \equiv (\beta - \alpha)/4$. Let $\{x_1,\ldots,x_n\}$ be an ϵ-approximation to S, and choose N in $\{1,\ldots,n\}$ such that $x_N > \sup\{x_1,\ldots,x_n\} - \epsilon$. Either $\alpha < x_N$ or $x_N < \alpha + 2\epsilon$. In the latter case, if s is any point of S, and i is chosen so that $|s - x_i| < \epsilon$, we have

$$s \leq x_i + \epsilon < x_N + 2\epsilon < \alpha + 4\epsilon = \beta,$$

so that β is an upper bound of S. Thus sup S (and similarly inf S) exists, by the constructive least-upper-bound principle (1.2). □

(4.5) Theorem *If f is a uniformly continuous map of a totally bounded space X into* $\mathbb{R}$, *then* inf $f \equiv$ inf $f(X)$ *and* sup $f \equiv$ sup $f(X)$ *exist.*

Proof. Apply (4.3) and (4.4). □

In particular, a uniformly continuous map f of a totally bounded space X into $\mathbb{R}$ is **bounded**, in the sense that $|f(x)| \leq c$ for some c in $\mathbb{R}$ and all x in X.

(4.6) Corollary *A totally bounded subset S of a metric space X is located.*

Proof. Let x_0 be any point of X. Then the map taking x to $\rho(x,x_0)$ is uniformly continuous, so that

$$\rho(x_0,S) \equiv \inf\{\rho(x,x_0) : x \in S\}$$

exists, by (4.5). □

The next theorem ensures that a totally bounded space has a plentiful supply of totally bounded subsets.

(4.7) Theorem *Let X be a totally bounded space, x_0 a point of X, and r a positive number. Then there exists a closed, totally bounded subset K of X such that $B(x_0,r) \subset K \subset B(x_0,8r)$.*

Proof. With $F_1 \equiv \{x_0\}$, we construct inductively a sequence (F_n) of finitely enumerable subsets of X such that

(i) $\rho(x,F_n) < 2^{-n+1}r$ for each x in $B(x_0,r)$,

(ii) $\rho(x,F_n) < 2^{-n+3}r$ for each x in F_{n+1}.

To this end, assume that $F_1,\ldots,F_n$ have been constructed with the appropriate properties. Let $\{x_1,\ldots,x_N\}$ be a $2^{-n}r$-approximation to X. Partition $\{1,\ldots,N\}$ into subsets A and B such that

$$\rho(x_i,F_n) < 2^{-n+3}r \text{ for } i \text{ in } A,$$
$$\rho(x_i,F_n) > 2^{-n+2}r \text{ for } i \text{ in } B.$$

Then, clearly, $F_{n+1} \equiv \{x_i : i \in A\}$ satisfies the appropriate instance of (ii). Let x be any point of $B(x_0,r)$. By the induction hypothesis, there exists y in F_n with $\rho(x,y) < 2^{-n+1}r$. Choosing i in $\{1,\ldots,N\}$ such that $\rho(x,x_i) < 2^{-n}r$, we have

$$\rho(x_i,F_n) \leq \rho(x_i,y) \leq \rho(x,x_i) + \rho(x,y) < 2^{-n+2}r.$$

Thus i cannot belong to B, and so $x_i \in F_{n+1}$. As $\rho(x,x_i) < 2^{-(n+1)+1}r$, the set F_{n+1} therefore satisfies the appropriate instance of (i). This completes the inductive construction.

Let K be the closure of $\bigcup_{n=1}^{\infty}F_n$ in X. We see from (i) that $B(x_0,r) \subset K$. On the other hand, if $m \geq n$ and $y \in F_m$, then by (ii) we can find points $y_m = y$, $y_{m-1} \in F_{m-1}$, $\ldots$, $y_n \in F_n$ such that $\rho(y_{i+1},y_i) < 2^{-i+3}r$ for $n \leq i \leq m-1$. Thus

(4.7.1) $$\rho(y,F_n) \leq \sum_{i=n}^{m-1} \rho(y_{i+1},y_i) < \sum_{i=n}^{\infty}2^{-i+3}r = 2^{-n+4}r.$$

It readily follows that F_n is a $2^{-n+5}r$-approximation to K; whence K is totally bounded. Finally, taking $n = 1$ in (4.7.1), we see that

$$\rho(y,x_0) = \rho(y,F_1) < 8r.$$

for each y in K. Thus $K \subset B(x_0,8r)$. This completes the proof. □

Note that in view of Lemma (4.5), the **diameter**

$$\operatorname{diam} X \equiv \sup\{\rho(x,y) : x,y \in X\}$$

exists for any totally bounded space X. Theorem (4.7) has the following corollary.

(4.8) Corollary *If X is a totally bounded space, then for each $\epsilon > 0$ there exist totally bounded sets $K_1,\ldots,K_n$, each of diameter less than ϵ, such that $X = \bigcup_{i=1}^{n}K_i$.*

Proof. Given $\epsilon > 0$, construct an $\epsilon/32$-approximation $\{x_1,\ldots,x_n\}$ to X. By (4.7), for each i in $\{1,\ldots,n\}$ there exists a closed, totally bounded set K_i such that $B(x_i,\epsilon/32) \subset K_i \subset B(x_i,\epsilon/4)$. Clearly, $X = \bigcup_{i=1}^{n} K_i$. Also, $\rho(x,y) < \epsilon/2$ for all x,y in K_i, so that diam $K_i \leq \epsilon/2 < \epsilon$. □

The standard notion of (pointwise) continuity of functions between metric spaces is not sufficiently powerful for most constructive purposes. Uniform continuity, however, is a concept of considerable constructive power (witness, for example, Proposition (4.5)). As will be made clear in Chapter 6, we have no hope of proving the classical uniform continuity theorem (a pointwise continuous function on a compact space is uniformly continuous); nor are we likely to find a constructive counterexample to that theorem.

The reader familiar with the classical theory of uniform spaces will appreciate that the following result is a constructive version of the theorem that the unique uniform structure U compatible with the given topology on a compact Hausdorff space X is induced by the U-uniformly continuous mappings of X into $\mathbb{R}$; the latter theorem is the key to the standard classical proof of the uniform continuity theorem in the general context of uniform spaces

(4.9) Proposition *Let h be a mapping of a metric space into a compact space X, such that $f \circ h$ is uniformly continuous for each uniformly continuous map $f: X \to \mathbb{R}$. Then h is uniformly continuous.*

Proof. Given any positive number ϵ, let $\{x_1,\ldots,x_n\}$ be an $\epsilon/3$-approximation to X. Define uniformly continuous maps $f_i: X \to \mathbb{R}$ by $f_i(x) \equiv \rho(x_i,x)$ for $i = 1,\ldots,n$. It suffices to show that if $|f_i(x) - f_i(x')|$ $< \epsilon/3$ for each i, then $\rho(x,x') < \epsilon$. Choose i such that $f_i(x) = \rho(x_i,x)$ $< \epsilon/3$, and therefore $f_i(x') < 2\epsilon/3$. Then

$$\rho(x,x') \leq \rho(x_i,x) + \rho(x_i,x') = f_i(x) + f_i(x') < \epsilon/3 + 2\epsilon/3 = \epsilon. \quad \square$$

It is instructive to compare this proof with a natural classical approach to Proposition (4.9). The classical proof we have in mind goes as follows. Suppose that h is not uniformly continuous on X. Then there exist (*sic*) $\alpha > 0$ and sequences (x_n), (y_n) in X, such that for each n, $\rho(x_n,y_n) < n^{-1}$ and $\rho(h(x_n),h(y_n)) \geq \alpha$. As X' is compact, there exists (*sic*) a subsequence $(x_{n(k)})_{k=1}^{\infty}$ of (x_n) such that $(h(x_{n(k)}))_{k=1}^{\infty}$ converges to a point z of X'. The uniform continuity of the map taking x to $\rho(h(x),z)$ on X now ensures that $(h(y_{n(k)}))_{k=1}^{\infty}$ also converges to z, whence $\rho(h(x_{n(k)}),h(y_{n(k)})) < \alpha$ for all sufficiently large k. Since thi contradicts our choice of (x_n) and (y_n), we conclude that h is uniformly continuous on X after all.

A nonvoid metric space X is **locally totally bounded** if each bounded subset of X is contained in a totally bounded subset; if, in addition, X is complete, then it is said to be **locally compact.** For example, the real line $\mathbb{R}$, the complex plane $\mathbb{C}$, and the euclidean spaces $\mathbb{R}^n$ and $\mathbb{C}^n$ are locally compact.

The following lemma prepares the way for our next theorem, which deals with certain fundamental properties of a locally totally bounded space.

(4.10) Lemma *Let L be a located subset of a metric space X, and T a totally bounded subset of X. Then there exists a totally bounded set S such that $T \cap L \subset S \subset L$.*

Proof. For each positive integer n, let T_n be a finite $1/n$-approximation to T. Write T_n as a union of finite sets A_n and B_n such that $\rho(t,L) < 3/n$ for all t in A_n, and $\rho(t,L) > 3/2n$ for all t in B_n. For each t in A_n, choose s_t^n in L such that $\rho(t,s_t^n) < 3/n$. Write $S_n \equiv \{s_t^n : t \in A_n\}$, and let S be the closure of $\bigcup_{n=1}^{\infty} S_n$ in L. To prove S totally bounded, consider positive integers m,n with $6m \leq n$, and any s in S_n. As $\rho(s,T) < 3/n$, we have

$$\rho(s,T_m) \; < \; 3/n + 1/m \; \leq \; 3/2m.$$

Choosing t in T_m such that $\rho(s,t) < 3/2m$, we see that t cannot belong to B_m and so belongs to A_m. Hence

$$\rho(s,s_t^m) \; \leq \; \rho(s,t) + \rho(t,s_t^m) \; < \; 3/2m + 3/m \; = \; 9/2m.$$

It now follows that $\bigcup_{k=1}^{6m} S_k$ is a finitely enumerable $9/2m$-approximation to $\bigcup_{n=1}^{\infty} S_n$; whence S is totally bounded.

If $x \in T \cap L$, and $n \in \mathbb{N}^+$, there exists t in T_n such that $\rho(x,t) < 1/n$. Then $\rho(t,L) < 1/n < 3/2n$, and so $t \in A_n$. Hence

$$\rho(x,s_t^n) \; < \; 1/n + 3/n \; = \; 4/n,$$

where $s_t^n \in S$. As x and n are arbitrary, and S is closed, $T \cap L \subset S$. □

(4.11) Theorem *Let Y be a nonvoid subset of a metric space X. Then*

(i) *if Y is locally totally bounded, it is located;*

(ii) *if X is locally totally bounded and Y is located, Y is locally totally bounded.*

Proof. Assume first that Y is locally totally bounded. Let $y_0 \in Y$ and $x \in X$. Construct a totally bounded subset T of Y containing all points y of Y such that $\rho(y,y_0) \leq 2\rho(x,y_0) + 1$. For each y in Y, either $\rho(x,y) > \rho(x,T)$ or $\rho(x,y) < \rho(x,y_0) + 1$. In the latter case,

$$\rho(y,y_0) \leq \rho(x,y) + \rho(x,y_0) \leq 2\rho(x,y_0) + 1,$$

so that $y \in T$, and therefore $\rho(x,y) \geq \rho(x,T)$. It follows that $\rho(x,Y)$ exists and equals $\rho(x,T)$. Thus Y is located. This proves (i); (ii) is a simple consequence of (4.10). □

The linear space $\mathbb{C}^n$ admits a number of natural norms, one of which is given by

$$\|(\lambda_1,\ldots,\lambda_n)\| = |\lambda_1| + \cdots + |\lambda_n|.$$

A normed linear space X is **finite-dimensional** if it contains a finitely enumerable set $\{e_1,\ldots,e_n\}$ - called a **metric basis** of X - such that the map $f:\mathbb{C}^n \to X$, defined by setting $f(\lambda_1,\ldots,\lambda_n) \equiv \Sigma_{i=1}^n \lambda_i e_i$, has a uniformly continuous inverse. In that case, $\{e_1,\ldots,e_n\}$ is a metrically finite set.

An important result, with at least one classical proof that readily adapts to a constructive one (**Bishop-Bridges**, Ch. 7, (2.3) and (2.6)), is the following:

(4.12) Proposition *The following are equivalent conditions on a normed linear space X:*

(i) *X is finite-dimensional;*

(ii) *the open unit ball of X is totally bounded;*

(iii) *the closed unit ball of X is compact;*

(iv) *X is locally totally bounded;*

(v) *X is locally compact.* □

5. Bounded Linear Maps

In this section we shall develop some elementary theory of bounded linear maps between normed linear spaces. As in the earlier sections, the material is chosen not so much for its intrinsic significance as for its peculiarly constructive content.

Let u be a linear map between normed linear spaces. We say that u is

(i) **bounded** if $u(B(0,1))$ is bounded,

(ii) **normable** if $\|u\| \equiv \sup\{\|y\| : y \in u(B(0,1))\}$ exists,

(iii) **compact** if $u(B(0,1))$ is totally bounded.

Every compact map is normable, by Theorem (4.5), every normable map is bounded, and every bounded map is uniformly continuous. If u is normable, then $\|u\|$ is called the **norm** of u. It is easy to give a Brouwerian example of a bounded linear functional - that is, a bounded linear map of a normed linear space into $\mathbb{C}$ - for which the norm does not exist.

If E is a subset of a vector space, and t is a scalar, then we let $tE \equiv \{tx : x \in E\}$. We say that E is **balanced** if $tE \subset E$ whenever $|t| \leq 1$. The closed unit ball in a normed linear space is balanced. Conversely, we have

(5.1) Proposition *Let B be a convex, balanced, closed subset of a linear space X, such that $\{|t| : x \in tB\}$ is nonempty and has an infimum $\|x\|_B$ for each x in X. If $x = 0$ whenever $\|x\|_B = 0$, then $\|\ \|_B$ is the unique norm on X with respect to which B is the closed unit ball.* □

The proof of Proposition (5.1) is left as an exercise (Problem 17). The norm $\|\ \|_B$ is called the **Minkowski norm** associated with B.

In the constructive development of the theory of bounded linear mappings it may be important to locate either the range or the **kernel** (null space),

$$\text{Ker}\, u \equiv \{x : u(x) = 0\},$$

of such a mapping u. For example, when u is a bounded operator on a Hilbert space H, the projection of H on the range of u exists if and only if the range is located in H. More general necessary and sufficient conditions for a located range seem hard to come by. However, with the aid of the next definition, we can give such conditions for a located kernel.

If V is a closed, located, linear subset of a normed linear space X, then the **quotient space** X/V is the linear space X taken with the **quotient norm**, defined by $\|x\|_{X/V} \equiv \rho(x,V)$, and the corresponding equality relation.

(5.2) Proposition *Let u be a bounded linear map of a normed linear space X onto a normed linear space Y. Then* Ker u *is located in X if and only if*

$\|y\|' \equiv \inf\{|t| : y \in tu(B(0,1))\}$ *exists for each* y *in* Y. *In that case,* u *is a norm-preserving isomorphism of* $X/\text{Ker } u$ *onto* $(Y, \| \ \|')$.

Proof. Observe that for each x in X, the following sets are equal:

$$\{t \in \mathbb{R}^+ : \|x - y\| < t \text{ for some } y \text{ in Ker } u\},$$

$$\{t \in \mathbb{R}^+ : u(x - tz) = 0 \text{ for some } z \text{ in } X \text{ with } \|z\| < 1\},$$

$$\{t \in \mathbb{R}^+ : u(x) \in tu(B(0,1))\}.$$

As $\rho(x, \text{Ker } u)$ is equal to the infimum of the first set, if either exists, and the infimum of the third set is equal to $\|u(x)\|'$, if either exists, the result follows from the definition of 'quotient space'. □

(5.3) Proposition *A nonzero bounded linear functional on a normed linear space* X *is normable if and only if its kernel is located in* X.

Proof. Let $u: X \to \mathbb{C}$ be a nonzero bounded linear functional. Suppose that u is normable, consider any complex number z, and let

$$A \equiv \{t : z \in tu(B(0,1))\}.$$

Then $\inf A = |z|/\|u\|$. As z is arbitrary, it follows from (5.2) that Ker u is located.

Conversely, if Ker u is located, then by (5.2),

$$\|u\| = 1/\inf\{|t| : 1 \in tu(B(0,1))\}$$

exists. □

It is easy to show that a nonzero bounded linear functional is compact if and only if it is normable. Thus Proposition (5.3) can be viewed as a special case of

(5.4) Theorem *A bounded linear mapping of a normed linear space* X *onto a finite-dimensional normed linear space is compact if and only if its kernel is located in* X.

For the proof of this theorem we require a special case of the open mapping theorem, and two other results.

(5.5) Lemma *Let* u *be a bounded linear map of a normed linear space* X *onto a finite-dimensional normed linear space* Y. *Then* $u(B(0,1))$ *is a neighbourhood of* 0 *in* Y.

Proof. If $Y = \{0\}$, this is trivial. Consider the case $n \geq 1$, and let $\{e_1, \ldots, e_n\}$ be a metric basis of Y. We may assume that for each i there exists x_i in $B(0,1)$ such that $u(x_i) = e_i$. Choose $\epsilon > 0$ so that if $y \equiv \Sigma_{i=1}^n \lambda_i e_i \in B(0,\epsilon)$, then $\Sigma_{i=1}^n |\lambda_i| < 1$. Given such y, we have $y = u(x)$, where $x \equiv \Sigma_{i=1}^n \lambda_i x_i$ and $\|x\| < 1$. Thus $B(0,\epsilon) \subset u(B(0,1))$. □

(5.6) Lemma *Let B be a totally bounded, balanced neighbourhood of 0 in a normed linear space Y containing a nonzero vector. Then* $\inf\{|t| : y \in tB\}$ *exists for each y in Y.*

Proof. Let $R \equiv \sup\{\|x\| : x \in B\}$, which is positive as B is a neighbourhood of 0 and Y is nontrivial. Then for each y in Y, $\inf\{|t| : y \in tB\} = \|y\|/R$. □

(5.7) Lemma *Let S be a finite-dimensional subspace of a normed linear space X, $\{e_1,\ldots,e_{n-1}\}$ a metric basis of S, and e_n a vector in X such that $\rho(e_n,S) > 0$. Then the set $\{e_1,\ldots,e_n\}$ is a metric basis of its span.*

Proof. It suffices to show that

$$\{(\lambda_1,\ldots,\lambda_n) : \|\Sigma_{i=1}^n \lambda_i e_i\| < 1\}$$

is bounded. As $\{e_1,\ldots,e_{n-1}\}$ is a metric basis of S, it suffices to find bounds on $|\lambda_n|$ and $\|\Sigma_{i=1}^{n-1}\lambda_i e_i\|$ when $\|\Sigma_{i=1}^n \lambda_i e_i\| < 1$. To this end, suppose that $\|\Sigma_{i=1}^n \lambda_i e_i\| < 1$. If $|\lambda_n| > 1/\rho(e_n,S)$, then

$$\left\|e_n + \frac{1}{\lambda_n}\sum_{i=1}^{n-1}\lambda_i e_i\right\| < \left|\frac{1}{\lambda_n}\right| < \rho(e_n,S),$$

which is a contradiction. So $|\lambda_n| \le 1/\rho(e_n,S)$ and

$$\left\|\sum_{i=1}^{n-1}\lambda_i e_i\right\| \le 1 + \|\lambda_n e_n\| \le 1 + \frac{\|e_n\|}{\rho(e_n,S)}. \quad □$$

(5.8) Lemma *Let X be a normed linear space. The set $\{e_1,\ldots,e_n\}$ is a metric basis of X if and only if (i) it spans X, and (ii) if $\lambda_1,\ldots,\lambda_n$ are scalars such that $\Sigma_{i=1}^n |\lambda_i| > 0$, then $\|\Sigma_{i=1}^n \lambda_i e_i\| > 0$.*

Proof. If $\{e_1,\ldots,e_n\}$ is a metric basis, then conditions (i) and (ii) clearly hold. Conversely, suppose those conditions obtain. If $n = 1$, then $\{e_1\}$ is easily seen to be a metric basis of X. If $n > 1$, assume that $\{e_1,\ldots,e_{n-1}\}$ is a metric basis of its span, S. In view of (5.7), it will suffice to prove that $\rho(e_n,S) > 0$. As S is finite-dimensional, it is complete and located; whence, by (3.3), there exists $s \in S$ such that if $\|e_n - s\| > 0$, then $\rho(e_n,S) > 0$. But $\|e_n - s\| > 0$, by (ii) and the assumption that $\{e_1,\ldots,e_{n-1}\}$ spans S; so $\rho(e_n,S) > 0$. □

We now return to the

Proof of Theorem (5.4). Let u be a bounded linear map of X onto a finite-dimensional normed linear space Y, and write $B \equiv u(B(0,1))$. If u is compact, then it follows from (5.5) and (5.6) that $\inf\{|t| : y \in tB\}$ exists for each y in Y; whence, by (5.2), Ker u is located. Conversely, if Ker u is located, then

$$\|u(x)\|' \equiv \rho(x, \text{Ker } u)$$

defines a norm, with a corresponding equality, on Y. It is easily seen that the open unit ball of $(Y, \| \ \|')$ is $u(B(0,1))$. Let $\{u(x_1),\ldots,u(x_n)\}$ be a metric basis for $(Y, \| \ \|)$. By (5.8), if $\lambda_1,\ldots,\lambda_n$ are scalars with $\Sigma_{i=1}^n |\lambda_i| > 0$, then $\|u(\Sigma_{i=1}^n \lambda_i x_i)\| > 0$; as u is continuous, it follows that

$$\|\Sigma_{i=1}^n \lambda_i u(x_i)\|' = \rho(\Sigma_{i=1}^n \lambda_i x_i, \text{Ker } u) > 0.$$

Hence $\{u(x_1),\ldots,u(x_n)\}$ is a metric basis of $(Y, \| \ \|')$, by (5.8). Therefore the identity maps from $(Y, \| \ \|)$ to $(Y, \| \ \|')$, and from $(Y, \| \ \|')$ to $(Y, \| \ \|)$, are uniformly continuous. By (4.12), $u(B(0,1))$ is totally bounded with respect to $\| \ \|'$; whence, by (4.3), it is totally bounded with respect to $\| \ \|$. Thus u is compact. □

In classical mathematics, every bounded linear mapping of a normed linear space onto a finite-dimensional normed linear space is compact. The reader is encouraged to construct a Brouwerian example of a bounded linear map u with range $\mathbb{R}^2$ such that $u(B(0,1))$ is not totally bounded (Problem 9).

The remainder of this section deals with the Hahn-Banach theorem. We will prove this theorem for real vector spaces, and then extend to the complex case. The first lemma shows how to extend a linear functional from a finite-dimensional subspace to a subspace with one more dimension.

(5.9) Lemma *Let X be a finite-dimensional subspace of a normed linear space Y over $\mathbb{R}$, and suppose that Y is spanned by $X \cup \{y_0\}$, where $\rho(y_0, X) > 0$. Let u be a nonzero normable linear functional on X. Then for each $\epsilon > 0$, there exists a normable linear functional v on Y such that $v = u$ on X, and $\|v\| \leq \|u\| + \epsilon$.*

Proof. We may assume that $\|u\| = 1$. For all $x, x' \in X$ we have

$$u(x) + u(x') = u(x + x') \leq \|x + x'\| \leq \|x - y_0\| + \|x' + y_0\|;$$

whence

$$u(x) - \|x - y_0\| \leq \|x' + y_0\| - u(x'),$$

and, *a fortiori*,

$$\varphi(x) \equiv u(x) - (1+\epsilon)\|x - y_0\| \leq (1+\epsilon)\|x' + y_0\| - u(x').$$

Choose $r > 0$ so that $\varphi(x) \leq -(1+\epsilon)\|y_0\|$ whenever $\|x\| \geq r$. Then

$$s \equiv \sup\{\varphi(x) : \|x\| \leq r\}$$

exists, by (4.12) and (4.5). By our choice of r, $s = \sup\{\varphi(x) : x \in X\}$. For each x in X we have

$$\text{(i)} \quad u(x) - s \leq (1+\epsilon)\|x - y_0\|,$$

$$\text{and} \quad \text{(ii)} \quad u(x) + s \leq (1+\epsilon)\|x + y_0\|.$$

Define a linear functional v on Y by setting

$$v(x + \alpha y_0) \equiv u(x) + \alpha s$$

for all $\alpha \in \mathbb{R}$. Clearly, v extends u. If $\alpha > 0$, then

$$v(x \pm \alpha y_0) = \alpha(u(\alpha^{-1}x) \pm s) \leq \alpha(1+\epsilon)\|\alpha^{-1}x \pm y_0\| = (1+\epsilon)\|x \pm \alpha y_0\|.$$

Hence $v(y) \leq (1+\epsilon)\|y\|$ for each y in Y; replacing y by $-y$, we have $|v(y)| \leq (1+\epsilon)\|y\|$.

It remains to show that v is normable. To this end, choose x_0 in X such that $u(x_0) = -s$. Then for each vector $x + \alpha y_0$ in Ker v, we have

$$x + \alpha y_0 = \alpha(x_0 + y_0) + (x - \alpha x_0),$$

where

$$u(x - \alpha x_0) = u(x) + \alpha s = v(x + \alpha y_0) = 0.$$

Hence Ker v is spanned by $\{x_0 + y_0\} \cup$ Ker u. Since u is normable and nonzero, Ker u is located in X, by (5.3); whence, by (4.11) and (4.12), Ker u is finite-dimensional. As

$$\rho(x_0 + y_0, \text{Ker } u) \geq \rho(y_0, X) > 0,$$

it follows from (5.7) that Ker v is finite-dimensional; whence it is located, by (4.12) and (4.11). Thus v is normable, by (5.3). □

To prove the real Hahn-Banach theorem, we iterate (5.9).

(5.10) The Real Hahn-Banach Theorem *Let u be a nonzero bounded linear functional on a linear subset Y of a separable normed linear space X over $\mathbb{R}$, such that the kernel of u is located in X. Then for each $\epsilon > 0$, there exists a normable linear functional v on X, extending u, such that*

$\|v\| \leq \|u\| + \epsilon$.

Proof. Let K be the kernel of u. By passing to X/K, we may assume that the kernel of u is 0, and that Y is one-dimensional. Let $x_1, x_2, \ldots$ be an enumeration of a dense subset of X such that $x_1 \in V$, $u(x_1) = 1$, and each element of the subset appears infinitely often. With $y_1 \equiv x_1$, construct a sequence $(y_1, y_2, \ldots)$ of vectors such that for each i, if Y_i is the space spanned by $y_1, \ldots, y_i$, then either

$$\text{(i)} \quad \rho(x_{i+1}, Y_i) < 1/i, \text{ in which case } y_{i+1} = 0,$$

$$\text{or} \quad \text{(ii)} \quad \rho(x_{i+1}, Y_i) > 0, \text{ in which case } y_{i+1} = x_{i+1}.$$

Then each Y_i is finite-dimensional, and $\cup_i Y_i$ is a dense subspace of X. By (5.9), for each i we can extend u from Y_i to Y_{i+1} and not increase its norm by more than $2^{-i}\epsilon$. Combining this chain of extensions, we obtain a normable linear functional v on a dense subspace of X, such that $\|v\| \leq \|u\| + \epsilon$. This functional extends uniquely to a linear functional on X with the same norm. □

The complex Hahn-Banach theorem follows immediately from (5.10) and the following lemma.

(5.11) Lemma *If X is a normed linear space over $\mathbb{C}$, then there is a one-one correspondence between real linear functionals u on X, and complex linear functionals u^* on X, given by*

$$u^*(x) = u(x) - iu(ix);$$

moreover, $\|u^*\| = \|u\|$.

Proof. If u^* is a complex linear functional, then $u^*(x) = u(x) + iv(x)$ for unique real linear functionals u and v. But

$$u(ix) + iv(ix) = u^*(ix) = iu^*(x) = -v(x) + iu(x),$$

so $v(x) = -u(ix)$. Conversely, if u is a real linear functional, then $u^*(x) \equiv u(x) - iu(ix)$ defines a linear functional u^* that respects scalar multiplication by real numbers; it is easily checked that $u^*(ix) = iu^*(x)$.

To prove that $\|u^*\| = \|u\|$, it is enough to show that $|u^*(x)| = \sup\{|u(\alpha x)| : |\alpha| = 1\}$. If $|\alpha| = 1$, then $|u^*(x)| = |u^*(\alpha x)| \geq |u(\alpha x)|$. On the other hand, if $\epsilon > 0$, then either $|u^*(x)| < |u(x)| + \epsilon$, or else $u^*(x) \neq 0$. In the latter case, if we let $\alpha \equiv |u^*(x)|/u^*(x)$, then $u^*(\alpha x) = \alpha u^*(x) = |u^*(x)| \in \mathbb{R}$, so $u^*(\alpha x) = u(\alpha x)$. Notice that in this case the supremum is achieved. □

6. Compactly Generated Banach Spaces

In this section we draw together some of the threads from Sections 1–5, in order to prove

(6.1) Theorem *A compactly generated Banach space is finite-dimensional.*

As we shall see, the proof of Theorem (6.1) is interesting (and provocative) in its own right; moreover, the theorem itself is not without application. But first we must clarify our terms.

To begin with, a **Banach space** is a complete, separable normed linear space. In CLASS, a Banach space need not be separable; our requirement of separability is not as restrictive as it may seem, as there are no interesting examples of constructively defined metric spaces that are not separable.

A subset G of a normed linear space X is a **generating set** for X if every element of X can be written in the form $\Sigma_{i=1}^{n}\lambda_i g_i$ for some $\lambda_1, \ldots, \lambda_n$ in the ground field and some $g_1, \ldots, g_n$ in G. If also G is compact, we say that X is **compactly generated.** Note that, as the reader may prove, a compactly generated normed linear space is separable. A finitely generated Banach space is compactly generated, because the closure of a finitely enumerable subset is compact.

A **subspace** of a normed linear space X is a closed located linear subset of X. Note that a finite-dimensional linear subset of a normed linear space is a subspace, in view of Proposition (4.12) and Theorem (4.11).

We now have a succession of lemmas that will lead us to Theorem (6.1).

(6.2) Lemma *Let X be a separable normed linear space, and Y a finite-dimensional subspace of X with nonvoid metric complement. Then there is a normable linear functional u on X such that $\|u\| = 1$ and $u(Y) = \{0\}$.*

Proof. Let x_0 be in the metric complement of Y. Define a linear functional v on $\mathbb{C}x_0 + Y$ by $v(x_0) \equiv 1$ and $v(Y) \equiv 0$. Then v is normable, with norm $1/\rho(x_0,Y)$. Using the Hahn-Banach theorem to extend v to a normable linear functional on X, we multiply the latter by a constant, to produce the required linear functional u of norm 1. □

The proofs of the next two lemmas are left as exercises

(Problems 21 and 12).

(6.3) Lemma *Let X be a normed linear space, K a totally bounded subset of X, and $\epsilon > 0$. Then there exists a finite-dimensional subspace Y of X such that $\rho(x,Y) < \epsilon$ for all x in K.* □

(6.4) Lemma *Let (X,ρ) be a complete metric space, and let a be a point of X such that $s \equiv \sup\{\rho(a,x) : x \in X\}$ exists. Then for each $r > 1$ there exists b in X such that $s \leq r\rho(a,b)$.* □

From a classical point of view, the following lemma is a triviality; however, its constructive proof requires some ingenuity. Note that if Y is a subspace of a Banach space X, then the quotient space X/Y is a Banach space (Problem 20).

(6.5) Lemma *Let X be a compactly generated Banach space, and Y a finite-dimensional subspace of X. Then either $X = Y$ or else $X - Y$ is nonvoid.*

Proof. Replacing X by the quotient space X/Y, we see that it is enough to prove that either $X = \{0\}$ or else there exists a nonzero element of X. Accordingly, let K be a compact generating set for X. We may assume that $0 \in K$. By (6.4), there exists b in K such that $\|x\| \leq 2\|b\|$ for all x in K. Define an increasing binary sequence (λ_n) such that

$$\lambda_n = 0 \Rightarrow \|b\| < 2/n^2,$$

and

$$\lambda_n = 1 \Rightarrow \|b\| > 1/n^2.$$

We may assume that $\lambda_1 = 0$. If $\lambda_n = 0$, set $x_n \equiv 0$; if $\lambda_n = 1$ and $\lambda_{n-1} = 0$, set $x_k \equiv nb$ for each integer $k \geq n$. Then

$$\|x_m - x_n\| < 2(n+1)/n^2$$

whenever $m > n$; so that (x_n) is a Cauchy sequence and therefore converges to a limit x_∞ in X. Choose elements $z_1,\ldots,z_N$ of K, and numbers $c_1, \ldots,c_N$, such that $x_\infty = \Sigma_{i=1}^N c_i z_i$. Choose also an integer $m > 2\Sigma_{i=1}^N |c_i|$. If $\lambda_m = 1$, then b is a nonzero element of X. Consider the case where $\lambda_m = 0$, and suppose that $\|b\| > 0$. Then $\lambda_n = 1$ for all sufficiently large n, so that $x_\infty = Mb$ for some $M > m$. Thus

$$M\|b\| \leq \Sigma_{i=1}^N |c_i| \|z_i\| \leq 2\Sigma_{i=1}^N |c_i| \|b\| < m\|b\|;$$

whence $M < m$, a contradiction. Hence, in fact, $b = 0$, so that $K = \{0\}$ and therefore $X = \{0\}$. □

(6.6) Lemma *Let X be a separable normed linear space such that for each finite-dimensional subspace Y of X, either $X = Y$ or else $X - Y$ is nonvoid. Let*

$$F \equiv \{x \in X : X \text{ is finite-dimensional}\},$$

let $K \subset X$ be compact, and let $\epsilon > 0$. Then there exists a in $(X - K) \cup F$ such that $\|a\| = \epsilon$.

Proof. Using (6.3), construct a finite-dimensional subspace Y of X such that $\rho(x,Y) < \epsilon/2$ for each x in K. Either $X = Y$, in which case $(X - K) \cup F = F = X$; or, as we may assume, $X - Y$ is nonvoid and, by (6.2), there exists a normable linear functional u on X with $\|u\| = 1$ and $u(Y) = \{0\}$. Choose a in X so that $\|a\| = \epsilon$ and $u(a) > 3\epsilon/4$. For each x in K there exists y in Y such that $\|x - y\| < \epsilon/2$; whence

$$\begin{aligned}\|a - x\| &\geq |u(a - x)| \geq |u(a - y)| - |u(x - y)| \\ &\geq |u(a)| - \|x - y\| > \epsilon/4.\end{aligned}$$

Thus $\rho(a,K) \geq \epsilon/4$, and therefore $a \in X - K \subset (X - K) \cup F$. □

(6.7) Lemma *Let X,F, and K be as in Lemma (6.6). Then $(X - K) \cup F$ is dense in X.*

Proof. Fix x in X and $\epsilon > 0$. Then the set

$$L \equiv \{y - x : y \in K\}$$

is compact, so that, by (6.6), there exists a in $(X - L) \cup F$ with $\|a\| = \epsilon$. If $a \in F$, then $(X - K) \cup F = F = X$, and there is nothing to prove. If $a \in X - L$, then for all y in K,

$$\|(a+x) - y\| = \|a - (y-x)\| \geq \rho(a,L) > 0,$$

so that $a+x \in X - K$. As $\|x - (a+x)\| = \|-a\| = \epsilon$, and as x,ϵ are arbitrary, the result follows. □

We can now give the

Proof of Theorem (6.1). Let G be a compact generating set for X. Then

$$K \equiv \{\alpha x : x \in G, |\alpha| \leq 1\}$$

is also a compact generating set for X. For each positive integer n, let K_n be the closure of the set of all sums of n elements of K. Then, as the reader may verify, K_n is compact, and $X = \bigcup_{n=1}^{\infty} K_n$. By (6.5) and (6.7), if

$$F \equiv \{x \in X : X \text{ is finite-dimensional}\},$$

then $(X - K_n) \cup F$ is dense in X; clearly, it is also open in X. Applying Baire's Theorem (1.3), construct a point x in $\bigcap_{n=1}^{\infty}((X - K_n) \cup F)$. Choose m in $\mathbb{N}^+$ such that $x \in K_m$. Then, as $x \in (X - K_m) \cup F$, x must belong to F. Thus F is nonvoid, and so X is finite-dimensional. □

The provocative aspect of this proof is its use of the set F to prove that X is finite-dimensional. To justify the claim that X is finite-dimensional, we must be able, at least in principle, to construct a metric basis of X. How does the proof that F is nonvoid translate into the construction of such a basis? The reader is invited to ponder this question.

We conclude with two applications of Theorem (6.1).

(6.8) Proposition *A locally compact normed linear space is finite-dimensional.*

Proof. Let X be a locally compact normed linear space. Then X is a Banach space. Any compact subset of X containing $B(0,1)$ is clearly a generating set for X. Hence X is finite-dimensional, by (6.1). □

(6.9) Proposition *If u is a compact linear map of a normed linear space V onto a Banach space X, then X is finite-dimensional.*

Proof. This follows from (6.1), since the closure of $u(B(0,1))$ in X is a compact generating set for X. □

PROBLEMS

1. Modify the proof of (1.2) so that only countable choice, not dependent choice, is involved: Construct a countable set A such that any u_n must lie in A, and decide in advance between the two conditions for each pair (n,s) in the countable set $\mathbb{N} \times A$.
2. Redo the proof of (2.1) so that the axiom of dependent choice is invoked explicitly.
3. Let a be a real number, and $\mathbb{R}a \equiv \{ax : x \in \mathbb{R}\}$. Prove that the following statements are equivalent:

(i) $a = 0$ or $a \neq 0$;

(ii) $\mathbb{R}a$ is located in $\mathbb{R}$;

(iii) $\mathbb{R}a$ is complete.

4. Construct a Brouwerian example of a metric space with a nonlocated ball.

5. Prove Theorem (2.5).

6. Prove that a located subset of a separable metric space is separable.

7. Show that Markov's Principle is equivalent to each of the following statements.

   (i) If A is a separable subset of a metric space X, then $\{x \in X : x \notin X - A\}$ is contained in the closure of A.

   (ii) If (C_n) is a sequence of closed colocated subsets of a separable complete metric space X, such that $X = \bigcup_{n=1}^{\infty} C_n$, then C_n has nonvoid interior for some n.

8. Construct a Brouwerian example of a located subset S of $\mathbb{R}^2$ such that $\mathbb{R}^2-S$ is not located.

9. Construct a Brouwerian example of a normable linear map u with range $\mathbb{R}^2$ such that $u(B(0,1))$ is not located in $\mathbb{R}^2$.

10. Let X be a complete metric space, and $f:X \to \mathbb{R}$ a pointwise continuous mapping such that $\inf f$ exists. Suppose that to each $\epsilon > 0$ there corresponds $\delta > 0$ such that $\rho(x,y) < \epsilon$ whenever $x,y \in X$ and $\max\{f(x),f(y)\} < \delta$. Show that there exists ξ in X such that if $f(\xi) > 0$, then $\inf f > 0$. (This problem was communicated to us by Peter Aczel.)

11. Prove **De Morgan's Rule for metric complements:** if (S_n) is a sequence of located subsets of a metric space X such that $S \equiv \bigcup_{n=1}^{\infty} S_n$ is complete and located, then $X - S = \bigcap_{n=1}^{\infty}(X - S_n)$.

12. Prove Lemma (6.4).

13. Let X be a compact space, and S a subset of X such that for each x in X there exists y in S such that if $n \in \mathbb{N}^+$ and $\rho(x,y) > 2^{-n-1}$, then $\rho(x,s) > 2^{-n+1}$ for all s in S. Prove that S is totally bounded. (*cf*. Lemma (3.3).)

14. Let $f:X \to \mathbb{R}$ be a continuous function on a compact metric space X,

and for each α in $\mathbb{R}$ define

$$X_\alpha \equiv \{x \in X : f(x) \leq \alpha\}.$$

Prove that there is a sequence (α_n) of real numbers greater than $m \equiv \inf f$ such that X_α is compact whenever $\alpha > m$ and $\alpha \neq \alpha_n$ for each n. (For each positive integer k write X as a finite union $\bigcup_{j=1}^{N(k)} X_{j,k}$ of compact sets, each of diameter less than k^{-1}. Define (α_n) to be an enumeration of the numbers $\inf\{f(x) : x \in X_{j,k}\}$ for all relevant values of j and k.)

15. Let h be a mapping of a compact space X into a metric space X', such that $f \circ h$ is uniformly continuous for each uniformly continuous map $f : X' \to \mathbb{R}$. Prove that h is pointwise continuous, that $h(X)$ is bounded, and that h is uniformly continuous if and only if $h(X)$ is totally bounded. Prove that if X' is locally compact, then h is uniformly continuous. (*cf*. Proposition (4.9))

16. Let E be a balanced subset of $\mathbb{C}$ containing a nonzero number. Prove that E is locally totally bounded if and only if $\inf\{|t|^{-1} : t \in E, t \neq 0\}$ exists.

17. Prove Proposition (5.1).

18. Construct a Brouwerian example of a linear subset V of $\mathbb{R}$ and a bounded linear functional u on V, such that (i) u is not normable, and (ii) u does not extend to a bounded linear functional on $\mathbb{R}$.

19. Show that the positive number ϵ is needed in the Hahn-Banach theorem, by constructing a Brouwerian example as follows. Let V be $\mathbb{R}^2$, with norm given by $\|(x,y)\| \equiv \sup\{|x|,|y|\}$, let a be a small real number, and let $p \equiv (1+a,1)$. Define $u : \mathbb{R}p \to \mathbb{R}$ by $u(rp) \equiv r\|p\|$, and suppose u extends to a linear functional v on V of norm 1. Let $e_1 \equiv (1+a,0)$ and $e_2 \equiv (0,1)$. Show that either $v(e_1) > 0$ or $v(e_2) > 0$; and that $v(e_2) = 0$ if $a > 0$, while $v(e_1) = 0$ if $a < 0$.

20. Show that if X is a Banach space, and Y is a subspace of X, then X/Y is a Banach space.

21. Prove Lemma (6.3).

22. A normed linear space X is **infinite-dimensional** if $X - V$ is

nonvoid for each finite-dimensional subspace V of X. Prove that if Y is a locally compact subset of an infinite-dimensional Banach space X, then $X - Y$ is dense in X. (Use (6.7), Baire's Theorem (1.3), and Problem 11.)

NOTES

One might argue that a sequence of real numbers is, by definition, a sequence of regular Cauchy sequences of rational numbers; thus there would be no second instance of countable choice in the proof of (1.1). Rephrased, this says that any function from $\mathbb{N}$ to $\mathbb{R}$ is a function from $\mathbb{N}$ to the set of regular Cauchy sequences of rational numbers. No one would make this last assertion with $\mathbb{N}$ replaced by $\mathbb{R}$, as it would imply LPO; so there appears to be something special about $\mathbb{N}$, which is exactly what countable choice says.

Classical proofs of Baire's theorem and the open mapping theorem are found in Chapter 5 of Rudin, *Real and Complex Analysis* (McGraw Hill, 1970). For more information on constructive versions of the open mapping theorem, see the forthcoming paper *Open and unopen mapping theorems*, by Bridges, Julian, and Mines; and Stolzenberg's unpublished paper *A critical analysis of Banach's open mapping theorem* (Northeastern University, 1971).

An elementary proof of Cantor's theorem (1.4) appears in Chapter 2 of **Bishop-Bridges**.

The completeness technique described after the proof of Theorem (2.4) first appeared in **Bishop** (Chapter 6, Lemma 7, p. 177). The technique has many interesting applications: for example, it is used at a key step in the proof of a constructive version of the fundamental theorem of approximation theory (**Bishop-Bridges**, Chapter 7, (2.8)).

The proof of Theorem (4.7) is taken from *Aspects of Constructivism*, an unpublished version of Errett Bishop's lectures at the Mathematics Symposium held at New Mexico State University in December, 1972.

The approach to Theorem (4.11), via Lemma (4.10), is neater and more natural than the standard one (found, for example, in **Bishop-Bridges**, Chap. 4, (6.2) and (6.3)).

Bishop defines a **continuous mapping** from a compact space X to a metric space Y to be one that is uniformly continuous; he defines a continuous mapping on a locally compact space X to be one that is either continuous on each compact subset of X, or, equivalently, uniformly continuous on each bounded subset of X.

The standard classical proof of the uniform continuity theorem in the context of uniform spaces is found in Volume I of Bourbaki's *General Topology* (Addison-Wesley, 1966, Chap. II, §4.1).

The locatedness of the range of a selfadjoint operator on a Hilbert space is discussed in *Operator ranges, integrable sets, and the functional calculus* (Houston J. Math. 11 (1985), 31-44).

Proposition (5.3) is proved in **Bishop-Bridges** (Chapter 7, (1.10)) without the help of Proposition (5.2). Theorem (5.4) first appears, with a slightly different proof, in *Bounded linear mappings of finite rank* (J. Functional Anal. 43 (1981), 143-148).

The Hahn-Banach theorem is proved as a consequence of the separation theorem in **Bishop** and **Bishop-Bridges.**

The material of Section 6 is drawn from *Compactly generated Banach spaces*, Archiv der Math. 36 (1981), 239-243.

Chapter 3. Russian Constructive Mathematics

In which the fundamentals of **RUSS** *are presented within a framework constructed from* **BISH** *and an axiom derived from Church's thesis. The notion of a programming system is introduced in Section 1, and is used in discussions of omniscience principles, continuity, and the intermediate value theorem. Section 3 deals with Specker's remarkable construction of a bounded increasing sequence of rational numbers that is eventually bounded away from any given (recursive) real number. The next section contains a strong counter-example to the Heine-Borel theorem, and a discussion of the Lebesgue measure of the unit interval and its compact subsets. The last two sections of the chapter deal with Ceitin's theorem on the continuity of real-valued functions.*

1. Programming Systems and Omniscience Principles

The Church-Markov-Turing notion of computability is justly hailed as a milestone in our understanding of effective procedures. However, its use as a general method for viewing mathematics from a constructive point of view has serious disadvantages. Recursive-function-theoretic formulations of theorems appear unnatural to the practising mathematician, and their proofs are often fraught with forbidding technical terminology that obscures the essential ideas. Indeed, an analyst might be forgiven for believing that the advantages to be gained by an understanding of the theorems of recursive analysis were not worth the effort necessary to achieve this understanding. This is unfortunate, because the essence of the basic theorems in recursive function theory can be obtained without a complicated technical formulation. We shall arrive at these theorems within the informal framework of BISH by appending an axiom that is a simple consequence of Church's thesis.

One of the insights of recursive function theory is that the proper study of computation involves *partial functions* rather than just functions.

Recall that a **partial function** from a set X to a set Y is a function f from a subset of X to Y. If $x \in X$, we say that $f(x)$ is **defined** if $x \in \operatorname{dom} f$, and that $f(x)$ is **undefined** if $x \in \operatorname{dom} f$ is impossible. A partial function f is a **restriction** of a partial function g if $\operatorname{dom} f \subset \operatorname{dom} g$ and $f(x) = g(x)$ for each x in $\operatorname{dom} f$. Two partial functions are **equal** if each is a restriction of the other. If $\operatorname{dom} f = X$ we say that f is **total.** Thus a total (partial) function from X to Y is simply a function from X to Y.

Partial functions from $\mathbb{N}$ to $\mathbb{N}$ arise from computer programs operating on inputs in $\mathbb{N}$. Imagine that each step of a program is executed in a fixed amount of time, and that the program signals when it stops executing, if ever, at which time the output can be viewed. We may associate with such a process a mapping A from $\mathbb{N} \times \mathbb{N}$ to $\mathbb{N} \cup \{-1\}$ by setting $A(x,n)$ equal to -1 if the program executes $n+1$ steps without stopping, and equal to the output if the program stops before completing $n+1$ steps. With this in mind, we say that a mapping A from $\mathbb{N} \times \mathbb{N}$ to $\mathbb{N} \cup \{-1\}$ is a **partial function algorithm** if $A(x,n+1) = A(x,n)$ whenever $A(x,n) \neq -1$. The **partial function f associated with A** has domain

$$\operatorname{dom} f = \{x \in \mathbb{N} : A(x,n) \neq -1 \text{ for some } n \text{ in } \mathbb{N}\};$$

if $x \in \operatorname{dom} f$, we define $f(x) \equiv A(x,n)$ for any n such that $A(x,n) \neq -1$. Such a partial function can be characterized by the property that its domain is countable. Before we prove this, note that a subset of $\mathbb{N}$ is countable if and only if it is the union of a sequence of finite sets.

(1.1) Lemma *A partial function from $\mathbb{N}$ to $\mathbb{N}$ has a countable domain if and only if it is equal to the partial function associated with some partial function algorithm.*

Proof. Let f be a partial function from $\mathbb{N}$ to $\mathbb{N}$. If f is associated with the partial function algorithm A, let $D(k) \equiv \{x \leq k : A(x,k) \neq -1\}$. Then $D(k)$ is a finite subset of $\mathbb{N}$ for each k, and $\operatorname{dom} f$ is the union of the sets $D(k)$. Thus $\operatorname{dom} f$ is countable.

Conversely, suppose that $\operatorname{dom} f$ is the union of a sequence $(D(k))$ of finite subsets of $\mathbb{N}$. Define $A(x,n) \equiv f(x)$ if $x \in D(k)$ for some $k \leq n$, and $A(x,n) \equiv -1$ otherwise. Then A is a partial function algorithm, and f is the associated partial function. □

The reader should prove that the composition of two partial functions $\mathbb{N}$ to $\mathbb{N}$ with countable domain is also a partial function with countable domain.

If L is a sufficiently powerful programming language, then we can write a program in L that enumerates all programs of L. Thus the set of partial functions arising from such programs is countable. If we believe that all functions worthy of the name are computable by the language L, then we are naturally led to accept the following axiom.

CPF **There is an enumeration $\varphi_0, \varphi_1, \ldots$ of the set of partial functions from $\mathbb{N}$ to $\mathbb{N}$ with countable domains.**

Such an enumeration of the set of partial functions with countable domains, together with a simultaneous enumeration of their domains, is called a (**universal**) **programming system.** For the rest of this chapter we shall assume CPF and consider a fixed programming system

$$\varphi_0, \varphi_1, \varphi_2, \ldots$$
$$D_0, D_1, D_2, \ldots,$$

where for each m, φ_m is a partial function from $\mathbb{N}$ to $\mathbb{N}$, and $(D_m(n))_{n=0}^{\infty}$ is a sequence of finite subsets of $\mathbb{N}$ such that $D_m(0) \subset D_m(1) \subset \ldots$, and $\mathrm{dom}\ \varphi_m = \bigcup_{n=0}^{\infty} D_m(n)$. For convenience, we take $D_m(n) \equiv \phi$ if $n < 0$.

We shall also adopt a fixed one-one correspondence between $\mathbb{N} \times \mathbb{N}$ and $\mathbb{N}$, called a **pairing function**, and denote the image of (x,y) in $\mathbb{N}$ by $\langle x,y\rangle$. This gives a one-one correspondence between $\mathbb{N}^n$ and $\mathbb{N}$ by the inductive definition $\langle x_1,\ldots,x_n\rangle \equiv \langle x_1,\langle x_2,\ldots,x_n\rangle\rangle$. Similarly, we shall adopt fixed enumerations of standard countable sets such as $\mathbb{Z}$ and $\mathbb{Q}$, and freely treat elements of such sets as if they were natural numbers. Thus we write $\varphi_m(x,y)$ when we consider φ_m as a function from $\mathbb{N} \times \mathbb{N}$ to $\mathbb{N}$, and $\varphi_m(x)$ when we consider φ_m as a mapping from $\mathbb{N}$ to $\mathbb{Q}$. This is very much in the spirit of RUSS, where everything is considered to be a natural number.

Until we discuss Ceitin's Theorem (Section 6 below), we will not employ any principles of RUSS other than CPF. The most important of the other principles is **Markov's principle:**

If (a_n) is a binary sequence such that $\neg\forall n(a_n = 0)$, then $\exists n(a_n = 1)$.

The argument for Markov's Principle is that we can find the desired n by successively computing $a_1, a_2, \ldots$, until we find n such that $a_n = 1$; we are guaranteed, by the hypothesis, that this procedure will terminate.

The argument against Markov's Principle is that we have no prior bound, in any sense, on how long it will take for the procedure to terminate.

Markov's Principle is false in many models of intuitionistic logic; indeed, according to Brouwer, it is incompatible with the intuitionistic notion of negation and the general idea of a choice sequence (which will be introduced in Chapter 5). Moreover, it is not used by the practitioners of BISH. For these reasons, we shall not use it without explicit mention.

It follows that we must maintain the distinction between *defined* and *not undefined*. Using our informal identification of a partial function algorithm with a computer program, we can interpret this distinction in the following way. If f is *not undefined* at x, then the program cannot run forever given the input x, but we may not have a prior bound for the number of steps it will take before it stops; if f is *defined* at x, then we have such a prior bound.

When we invoke CPF in the framework of BISH, we are, in effect, working in RUSS. Note that, in spite of CPF, the set of partial function *algorithms* is not countable, because any enumeration of them could be trivially converted to an enumeration of all (total) functions from $\mathbb{N}$ to $\mathbb{N}$: a standard diagonal argument shows that the latter enumeration is not possible.

The problem of determining whether or not $\varphi_m(x)$ is defined is called the **halting problem**. To be more precise, we may state the halting problem in the following way:

Find a function $h : \mathbb{N} \times \mathbb{N} \to \{0,1\}$ *such that for each pair* (m,x) *in* $\mathbb{N} \times \mathbb{N}$, $\varphi_m(x)$ *is defined if and only if* $h(m,x) = 1$.

The halting problem is clearly related to LPO (see Corollary 1.4); indeed, it is not difficult to show that the existence of such a function h is equivalent to LPO. This suggests that the halting problem cannot be solved. We can state this more positively as follows.

(1.2) Lemma *For each (total) function* $g:\mathbb{N} \to \{0,1\}$, *there exists* m *in* $\mathbb{N}$ *such that* $g(m) = 0$ *if and only if* $\varphi_m(m)$ *is defined. Hence there is no function* $g:\mathbb{N} \to \{0,1\}$ *with the property that* $g(m) = 1$ *if and only if* $\varphi_m(m)$ *is defined (or not undefined).*

Proof. It is readily seen that $g^{-1}(0) = \bigcup_{n=0}^{\infty}(g^{-1}(0) \cap \{0,1,\dots,n\})$ is countable. Choose m such that φ_m is a partial function with domain $g^{-1}(0)$. Then $\varphi_m(m)$ is defined if and only if $g(m) = 0$. □

(1.3) Corollary *The halting problem cannot be solved.*

Proof. Suppose h were a mapping as in the statement of the halting problem. Then $g(m) \equiv h(m,m)$ defines a function from $\mathbb{N}$ to $\{0,1\}$ such that $g(m) = 1$ if and only if $\varphi_m(m)$ is defined. This contradicts (1.2). □

(1.4) Corollary LPO *is false.*

Proof. Assume that LPO obtains. Then there is a function λ, with domain the set S of all binary sequences, such that $\lambda(a) = 1$ if $a_n = 1$ for some n, and $\lambda(a) = 0$ if $a_n = 0$ for all n. Let μ be the function from $\mathbb{N}$ to S such that $\mu(m)_k = 1$ if and only if $m \in D_m(k)$, and let $g \equiv \lambda\circ\mu$. Then $g(m) = 1$ if and only if $\varphi_m(m)$ is defined, which contradicts (1.2). □

As LPO is demonstrably false in RUSS, Brouwerian counterexamples involving LPO become counterexamples in the ordinary sense. In particular, the full axiom of choice is false in RUSS (see Section 4 of Chapter 1). What about Brouwerian counterexamples involving LLPO? The following theorem shows that LLPO, in conjunction with countable choice, is false.

(1.5) Theorem *There is a total function $F:\mathbb{N} \times \mathbb{N} \to \{0,1\}$ such that*

(i) *for each m there is at most one n such that $F(m,n) = 1$;*

(ii) *if f is a total function from $\mathbb{N}$ to $\{0,1\}$, then there exist m and k in $\mathbb{N}$ such that $F(m, 2k+f(m)) = 1$.*

Proof. Set

$$F(m,n) = 1 \quad \text{if either } n = 2k,\ m \in D_m(k)\backslash D_m(k-1), \text{ and } \varphi_m(m) = 0,$$
$$\text{or } n = 2k+1,\ m \in D_m(k)\backslash D_n(k-1), \text{ and } \varphi_m(m) > 0,$$
$$= 0 \quad \text{otherwise.}$$

Clearly, F satisfies (i). Given a total function $f:\mathbb{N} \to \{0,1\}$, choose m,k such that $f = \varphi_m$ and $m \in D_m(k)\backslash D_m(k-1)$. If $f(m) = 1$, then $\varphi_m(m) = 1$, and so $F(m,2k+1) = 1$; while if $f(m) = 0$, then $\varphi_m(m) = 0$, and so $F(m,2k) = 1$. Thus F satisfies (ii). □

(1.6) Corollary LLPO *is false.*

Proof. Assume that LLPO obtains, and let F be as in (1.5). Then for each m, either $F(m,2k) = 0$ for all k, or $F(m,2k+1) = 0$ for all k. By countable choice, there exists a function f from $\mathbb{N}$ to $\{0,1\}$ such that if $f(m) = 0$, then $F(m,2k) = 0$ for all k, and if $f(m) = 1$, then $F(m,2k+1) = 0$ for all k. This contradicts (1.5) □

Countable choice in RUSS can usually be avoided by postulating a total function $c : \mathbb{N} \times \mathbb{N} \to \mathbb{N}$ such that $\varphi_i \circ \varphi_j = \varphi_{c(i,j)}$ for all i and j. Such a **composition function** c exists by countable choice, but in the context of programming languages there is normally a natural construction. One consequence of the existence of a composition function is the **s-m-n theorem:**

(1.7) Theorem *There is a total function s such that for all i, m, and n,*

$$\varphi_{s(i,m,x_1,\ldots,x_m)}(y_1,\ldots,y_n) = \varphi_i(x_1,\ldots,x_m,y_1,\ldots y_n).$$

Proof. Given m in $\mathbb{N}$, choose k in $\mathbb{N}$ such that $\varphi_k(m,x,y) = \langle x_1,\ldots,x_m,y\rangle$ for all x and y, where $x = \langle x_1, \ldots,x_m\rangle$ and $y = \langle y_1, \ldots,y_n\rangle$. Then it suffices to find s so that

$$\varphi_{s(i,m,x)}(y) = \varphi_i(\varphi_k(m,x,y)).$$

Choose p and q such that $\varphi_p(y) = \langle 0,y\rangle$ and $\varphi_q(x,y) = \langle x+1,y\rangle$ for all x and y. Let R be the total function defined inductively by $R(0) \equiv p$ and $R(x + 1) \equiv c(q,R(x))$, where c is the composition function. Then for all x,y, and z we have

$$\varphi_{R(x)}(y) = \langle x,y\rangle$$

and

$$\varphi_{R(x)}(\varphi_{R(y)}(z)) = \varphi_{R(x)}\langle y,z\rangle = \langle x,y,z\rangle.$$

The required function s is now given by

$$s(i,m,x) \equiv c(i,c(k,c(R(m),R(x)))).$$ □

2. Continuity and intermediate values

So far, we have used CPF mainly to establish negative results; now we prove some positive theorems about continuity. We begin with a lemma showing that functions on $\mathbb{R}$ cannot be sequentially discontinuous.

An **operation** from a set X to a set Y is a function g from X to $\mathcal{P}(Y)$ such that $g(x)$ is nonempty for each x in X.

(2.1) Lemma *If X is a closed subset of $\mathbb{R}$, and (a_n) a sequence in X converging to a limit ℓ, then there is no operation g from X to $\{0,1\}$ such that $g(a_n) = \{1\}$ and $g(\ell) = \{0\}$.*

Proof. Suppose g is such an operation, and define $f:\mathbb{N} \times \mathbb{N} \to \{0,1\}$ by setting $f(m,n) \equiv 1$ if and only if $m \notin D_m(n)$. Given $\epsilon > 0$, choose N in $\mathbb{N}$ so that $|a_j - a_k| < \epsilon$ whenever $j \geq k \geq N$. Then for such j,k, and for all m, we have

$$|\Sigma_{n=k}^{j} f(m,n)(a_n - a_{n-1})| < \epsilon.$$

As $\epsilon > 0$ is arbitrary, it follows that for each m, the series

$$a_0 + \Sigma_{n=1}^{\infty} f(m,n)(a_n - a_{n-1})$$

converges to a sum r_m in X. If $\varphi_m(m)$ is defined, then $r_m = a_n$ for some n; whence $g(r_m) = \{1\}$. Thus if $0 \in g(r_m)$, then $\varphi_m(m)$ is undefined. Conversely, if $\varphi_m(m)$ is undefined, then $r_m = \ell$; whence $g(r_m) = \{0\}$. Since g is an operation from X to $\{0,1\}$, either $0 \in g(r_m)$, in which case $g(r_m) = \{0\}$; or $1 \in g(r_m)$, in which case $0 \notin g(r_m)$, and therefore $g(r_m) = \{1\}$. Therefore we can define a total function $G:\mathbb{N} \to \{0,1\}$ by $G(m) \in g(r_m)$ for each m. Then $G(m) = 1$ if and only if $\varphi_m(m)$ is not undefined; this contradicts (1.2). □

Our first result about continuity concerns integer-valued functions. An easy interval-halving argument in BISH shows that an integer-valued function on $\mathbb{R}$ is continuous if and only if it is constant; so the following theorem, which depends on CPF, says that integer-valued functions on $\mathbb{R}$ are continuous.

(2.2) Theorem *Every integer-valued function on $\mathbb{R}$ is constant.*

Proof. It suffices to consider a function $f:\mathbb{R} \to \{0,1\}$. Suppose there exist real numbers a and b such that $f(a) = 0$ and $f(b) = 1$. Using a standard approximate interval-halving argument, we can construct Cauchy sequences (a_n) and (b_n) of real numbers such that

(i) $f(a_n) = 0$ and $f(b_n) = 1$ for $n \geq 1$,

(ii) $a_n - b_n \to 0$ as $n \to \infty$.

By (ii), the two sequences converge to a common limit ℓ in $\mathbb{R}$. We may assume that $f(\ell) = 1$. Then, setting $g(x) = \{f(x)\}$, we obtain a contradiction to (2.1). Hence f is constant. □

Following Markov, we next prove a weak form of continuity - namely, **sequential nondiscontinuity** - for arbitrary functions on $\mathbb{R}$.

(2.3) Theorem *If $X \subset \mathbb{R}$ is a closed set, (a_n) is a sequence in X converging to a limit ℓ, and $f:X \to \mathbb{R}$ is a function such that $|f(a_n) - f(\ell)| \geq \delta$ for all n, then $\delta \leq 0$.*

Proof. Suppose $\delta > 0$, and define an operation g from X to $\{0,1\}$ by

$$0 \in g(x) \text{ if and only if } |f(x) - f(\ell)| < \delta,$$
$$1 \in g(x) \text{ if and only if } |f(x) - f(\ell)| > 0.$$

Then $g(\ell) = \{0\}$, and $g(a_n) = \{1\}$ for each n. As this contradicts (2.1), we have $\delta \leq 0$. □

A remarkable result of Ceitin states that, in the full context of RUSS (that is, with certain principles over and above CPF), every real-valued function on an interval is pointwise continuous. We shall prove Ceitin's theorem in Section 6 below. In the meantime, we turn to the **intermediate value theorem:**

> *If $f:[a,b] \to \mathbb{R}$ is a uniformly continuous function, and t is a real number such that $f(a) < t < f(b)$, then there exists s in (a,b) such that $f(s) = t$.*

We shall show that the intermediate value theorem is equivalent to LLPO in BISH; whence, by Corollary (1.6), it is false in RUSS.

(2.4) Theorem *The intermediate value theorem is equivalent to* LLPO.

Proof. Suppose the intermediate value theorem holds. Given t in $[-1,1]$, we shall show that either $t \in [-1,0]$ or $t \in [0,1]$. Let f be the uniformly continuous function on $[-1,1]$ such that

$$\begin{aligned} f(x) &= x + 1/2 && \text{for} \quad -1 \leq x \leq -1/2, \\ f(x) &= 0 && \text{for} \quad -1/2 \leq x \leq 1/2, \\ f(x) &= x - 1/2 && \text{for} \quad 1/2 \leq x \leq 1. \end{aligned}$$

The intermediate value theorem produces s in $(-1,1)$ such that $f(x) = t$. If $s < 1/2$, then $t \in [-1,0]$; while if $s > -1/2$, then $t \in [0,1]$. Thus LLPO holds.

Conversely, if LLPO holds, and f satisfies the hypotheses of the intermediate value theorem, then an interval-halving argument yields the required point s in (a,b). The details are left to the reader. □

We say that a function f **approximates intermediate values** if, whenever $a \leq b$, $f(a) < t < f(b)$, and $\epsilon > 0$, there exists s in $[a,b]$ such that $|f(s) - t| < \epsilon$. A standard theorem of BISH states that uniformly continuous functions approximate intermediate values; indeed, in the presence of countable choice, only *pointwise* continuity is required. In view of Ceitin's theorem, *arbitrary* functions approximate intermediate values in RUSS. We shall show that CPF suffices for this result, and that, for sufficiently well-behaved functions, the intermediate values are attained.

A function $f:[a,b] \to \mathbb{R}$ is said to be **locally nonconstant** if for each x in $[a,b]$ and each $\epsilon > 0$, there exists x' in $[a,b]$ such that $f(x) \neq f(x')$ and $|x - x'| < \epsilon$.

(2.5) Theorem *Let f be a function from the closed interval $[a,b]$ to $\mathbb{R}$, and suppose that $f(a) < t < f(b)$. Then for each $\epsilon > 0$ there exists s in $[a,b]$ such that $|f(s) - t| < \epsilon$; moreover, if f is locally nonconstant, then s can be found so that $f(s) = t$.*

Proof. Fix $\epsilon > 0$. Using dependent choice and approximate interval-halving, we can construct sequences (a_n) and (b_n) such that for each n,

$$a = a_1 \leq a_2 \leq \cdots \leq a_n \leq b_n \leq \cdots \leq b_2 \leq b_1 = b,$$

$$b_n - a_n \leq (2/3)^{n-1}(b - a),$$

and either

(i) $f(a_n) < t < f(b_n)$,

or

(ii) $a_n = b_n$ and $|f(a_n) - y| < \epsilon/3$.

Moreover, if f is locally nonconstant, we can arrange that (i) holds for each n, and that the construction is independent of ϵ. Then (a_n) and (b_n) are Cauchy sequences with a common limit s in $[a,b]$. Either $|f(s) - t| < \epsilon$ or $|f(s) - t| > \epsilon/2$; we shall rule out the latter possibility. This will also cover the locally nonconstant case, in which the construction does not depend on ϵ.

By symmetry, we may assume that $f(s) > t + \epsilon/2$. For each n, either (i) holds, and

$$f(s) - f(a_n) > f(s) - t > \epsilon/2;$$

or else (ii) holds, so that

$$|f(s) - f(a_n)| \geq |f(s) - t| - |f(a_n) - t|$$
$$\geq \epsilon/2 - \epsilon/3 = \epsilon/6.$$

Thus $|f(s) - f(a_n)| \geq \epsilon/6$ for all n, which contradicts (2.3). □

3. Specker's Sequence

We have already commented that LPO can be derived from the assertion that every bounded increasing sequence of real numbers converges; the latter assertion is therefore false in RUSS. In a remarkable paper, Specker showed that a much sharper statement holds in RUSS: there is a bounded increasing sequence of rational numbers which is eventually bounded away from any given real number.

To prove Specker's theorem, first recall that the **Cantor set** is the subset

$$C \equiv \{\Sigma_{n=1}^{\infty} c_n 3^{-n} : c_n \in \{0,2\} \text{ for each } n\}$$

of the unit interval [0,1]. It is well known that C is a closed, and therefore complete, set of real numbers. Note that if a and b are two numbers in C which differ in the mth place of their ternary expansions, then $|a - b| \geq 3^{-m}$. Also, for each positive integer n, the finite set

$$\{\Sigma_{k=1}^{n} c_k 3^{-k} : c_k \in \{0,2\} \quad \text{for} \quad 1 \leq k \leq m\}$$

is a 3^{-n}-approximation to C; thus C is totally bounded, and therefore located.

We now prove **Specker's Theorem:**

(3.1) Theorem *There exists an increasing sequence* (r_n) *of rational numbers in the Cantor set such that for each* x *in* $\mathbb{R}$*, there exist* N *in* $\mathbb{N}$ *and* $\delta > 0$ *satisfying* $|x - r_n| \geq \delta$ *whenever* $n \geq N$.

Proof. For each positive integer n, let $r_n = \Sigma_{i=1}^{n} s_n(m)3^{-m}$, where

$$s_n(m) = 2 \quad \text{if } m \in D_m(n) \text{ and } \varphi_m(m) = 0,$$
$$= 0 \quad \text{otherwise.}$$

To begin with, let x be an arbitrary number in the Cantor set C. There exists m in $\mathbb{N}$ such that $\varphi_m : \mathbb{N} \to \{0,2\}$ is a total function and

$$x = \Sigma_{k=1}^{\infty} \varphi_m(k)3^{-k}.$$

Let $\delta \equiv 3^{-m}$, and choose $N \geq m$ so large that $m \in D_m(N)$. Then since $\varphi_m(m) \in \{0,2\}$, $\varphi_m(m) = 2 - s_n(m)$ if $n \geq N$. Hence $|x - r_n| \geq 3^{-m}$ for all

$n \geq N$. This completes the proof when $x \in C$.

Now consider an arbitrary real number x. Define an increasing binary sequence λ so that for each n,

$$\lambda(n) = 0 \Rightarrow \rho(x,C) < 1/n,$$
$$\lambda(n) = 1 \Rightarrow \rho(x,C) > 1/(n+1).$$

Let c be any point of C. If $\lambda(1) = 1$, set $a_n \equiv c$ for all n in $\mathbb{N}$; if $\lambda(n) = 0$, choose a_n in C with $|x - a_n| < 1/n$; if $\lambda(1) = 0$ and $\lambda(n) = 1$, set $a_n \equiv a_{n-1}$. Then (a_n) is a Cauchy sequence in C: in fact, $|a_m - a_n| \leq 2/n$ whenever $m \geq n$. Since C is complete, (a_n) converges to a limit a in C. By the first part of the proof, there exist N in $\mathbb{N}$ and $\delta' > 0$ such that $|a - r_n| \geq \delta'$ whenever $n \geq N$. Write

$$\delta \equiv \min\{1/(m+1), \delta' - |x - a|\}.$$

Either $|x - a| < \delta'$, in which case

$$|x - r_n| \geq |a - r_n| - |x - a| \geq \delta$$

whenever $n \geq N$; or else $|x - a| > 0$. In the latter case, choosing m such that $|x - a_m| > 1/m$, we see that $\lambda(m) \neq 0$; whence $\lambda(m) = 1$, and so $\rho(x,C) > 1/(m+1)$. Then $|x - r_n| \geq \rho(x,C) \geq \delta$ for all n. □

(3.2) Corollary *There exists a sequence $I_1, I_2, \ldots$ of disjoint intervals with rational endpoints contained in [0,1], such that*

(i) *the lengths of the intervals are positive and converge to 0;*

(ii) *for each real number x there exist N in $\mathbb{N}$ and $\delta > 0$ such that $\rho(x, I_n) > \delta$ whenever $n \geq N$.*

Proof. Let $r_1, r_2, \ldots$ be the sequence of Theorem (3.1). As $|r_m - r_n| > 0$ for any given m and all sufficiently large n, by passing to a subsequence we may assume that $0 < r_1 < r_2 < \ldots$. Construct a sequence (e_n) of positive rational numbers converging to 0, such that for each n,

$$e_n < \min\{r_{n+1} - r_n, r_n - r_{n-1}\}.$$

Then we need only set $I_n \equiv [r_n - e_n, r_n + e_n]$ for each n. □

Corollary (3.2) leads to many curious results in RUSS; for example, we have

(3.3) Theorem *There exists a pointwise continuous function from [0,1] onto (0,1) that is not uniformly continuous.*

Proof. For each n in $\mathbb{N}^+$, let I_n be as in (3.2), and let $t_n(x)$ be a continuous function from [0,1] to $\mathbb{R}$ that maps I_n onto $[-1+n^{-1},1-n^{-1}]$ and vanishes outside I_n. Define $f(x) \equiv \sum_{n=1}^{\infty} t_n(x)$. Because of (3.2), the function f is well-defined and pointwise continuous. Clearly, f maps [0,1] onto (-1,1), so that $(1+f)/2$ maps [0,1] onto (0,1). The function $(1+f)/2$ is not uniformly continuous, because it maps arbitrarily small intervals I_n onto intervals of length greater than 1/2. □

A much more remarkable theorem states that we can find a *uniformly* continuous map from [0,1] onto (0,1). We shall establish this in Chapter 6.

4. The Heine-Borel Theorem

Bearing in mind the classical proofs of the uniform continuity theorem, we see from Theorem (3.3) that the Heine-Borel theorem of classical analysis must be false in RUSS. Our experience of RUSS leads us to wonder if there might be a result which asserts more than just the contradictoriness of the Heine-Borel theorem. Indeed there is.

We write $|I|$ for the length of an interval in $\mathbb{R}$.

(4.1) Theorem *For each $\epsilon > 0$ there exists a sequence $(I_k)_{k=1}^{\infty}$ of bounded open intervals in $\mathbb{R}$ such that (i) $\mathbb{R} \subset \bigcup_{k=1}^{\infty} I_k$, and (ii) $\sum_{k=1}^{n} |I_k| < \epsilon$ for each n in $\mathbb{N}^+$.*

Proof. For the purposes of this proof, we consider $\varphi_0, \varphi_1, \ldots$ to be an enumeration of the set of partial functions from $\mathbb{N}$ to $\mathbb{Q}$.

Given $\epsilon > 0$, choose a positive integer $N > 4/\epsilon$. For each pair (m,n) of positive integers, set

$$J_{m,n} \equiv (\varphi_m(2^m N) - 2^{-m+1}N^{-1}, \varphi_m(2^m N) + 2^{-m+1}N^{-1})$$

if $2^m N \in D_m(n) \setminus D_m(n-1)$, and set $J_{m,n} \equiv \phi$ otherwise. Then if $x \equiv (x_n)_{n=1}^{\infty} \in \mathbb{R}$, we can find m such that $x_n = \varphi_m(n-1)$ for each $n \in \mathbb{N}^+$. As

$$|x - \varphi_m(2^m N)| \leq (2^m N + 1)^{-1} < 2^{-m+1}N^{-1},$$

we see that x belongs to the nonempty open interval $J_{m,n}$, where n is the unique integer such that $2^m N \in D_m(n) \setminus D_m(n-1)$. Hence $\mathbb{R} \subset \bigcup_{m,n=1}^{\infty} J_{m,n}$. As it is decidable whether or not $J_{m,n}$ is empty for any given m and n, we can find an enumeration $I_1, I_2, \ldots$ of the nonvoid terms of the double sequence $(J_{m,n})$. Then (i) clearly holds. Also, for each positive integer

n we have

$$\Sigma_{k=1}^{n}|I_k| \leq \Sigma_{m=1}^{\infty}2^{-m+2}N^{-1} = 4N^{-1} < \epsilon. \quad \square$$

The proof of the following corollary is left as an exercise.

(4.2) Corollary *Let (I_n) be as in Theorem (4.1); then $\mathbb{R} - \bigcup_{k=1}^{n}I_k$ is nonempty for each n in $\mathbb{N}^+$.* $\square$

At first sight, Theorem (4.1) appears to destroy any hope for a satisfactory development of measure theory in RUSS, as it suggests that $\mathbb{R}$ will have Lebesgue measure equal to 0. On the other hand, our next theorem shows that the interval [0,1] should have Lebesgue measure at least 1.

If I and J are two bounded intervals of $\mathbb{R}$, we define the **length** of their intersection to be the nonnegative number

$$|I \cap J| \equiv \max\{0, \min\{b,d\} - \max\{a,c\}\},$$

where $\overline{I} = [a,b]$ and $\overline{J} = [c,d]$. Note that if $I \cap J$ is nonempty, then $|I \cap J|$ is the length of $I \cap J$ in the usual sense.

(4.3) Lemma *If I and J are compact intervals with $|I \cap J| < |I|$, then $I - J$ is nonvoid.*

Proof. Let $I \equiv [a,b]$ and $J \equiv [c,d]$. By hypothesis

$$\min\{b,d\} - \max\{a,c\} < b - a,$$

and therefore

$$(b - \min\{b,d\}) + (\max\{a,c\} - a) > 0.$$

It follows that either $b > \min\{b,d\}$ or $\max\{a,c\} > a$. In the first case, $b \in I - J$, and in the second, $a \in I - J$. $\square$

(4.4) Lemma *Let a,b be real numbers with $a < b$, and let J be a bounded interval in $\mathbb{R}$. Then*

$$|[a,\tfrac{1}{2}(a+b)] \cap J| + |[\tfrac{1}{2}(a+b),b] \cap J| \leq |J|.$$

Proof. The proof is left as an exercise. $\square$

The proof of our next theorem, which holds within BISH, is due to Newcomb Greenleaf.

(4.5) Theorem *Let I be a compact interval, and $(J_n)_{n=1}^{\infty}$ a sequence of bounded open intervals, such that $\Sigma_{n=1}^{\infty}|J_n|$ converges to a sum less than*

$|I|$. *Then* $\bigcap_{n=1}^{\infty}(I - J_n)$ *is nonvoid.*

Proof. Write $I_1 \equiv I \equiv [a,b]$. Let

$$0 < \alpha < |I| - \Sigma_{n=1}^{\infty}|J_n|,$$

and choose a positive integer N so that $\Sigma_{n=N+1}^{\infty}|J_n| < \alpha/2$. Then, by (4.4),

$$\Sigma_{n=1}^{N}|[a,\tfrac{1}{2}(a+b)] \cap J_n| + \Sigma_{n=1}^{N}|[\tfrac{1}{2}(a+b),b] \cap J_n|$$
$$\leq \Sigma_{n=1}^{N}|J_n| < |I| - \alpha.$$

Hence either

$$\Sigma_{n=1}^{N}|[a,\tfrac{1}{2}(a+b)] \cap J_n| < \tfrac{1}{2}(|I| - \alpha)$$

or

$$\Sigma_{n=1}^{N}|[\tfrac{1}{2}(a+b),b] \cap J_n| < \tfrac{1}{2}(|I| - \alpha).$$

In the first case, set $I_2 \equiv [a,\frac{1}{2}(a+b)]$; in the second, set $I_2 \equiv [\frac{1}{2}(a+b),b]$. In each case we have

$$\Sigma_{n=1}^{\infty}|I_2 \cap J_n| \leq \Sigma_{n=1}^{N}|I_2 \cap J_n| + \Sigma_{n=N+1}^{\infty}|J_n|$$
$$< \tfrac{1}{2}(|I| - \alpha) + \alpha/2 = |I_2|.$$

Carrying on in this manner, we construct inductively a sequence $(I_m)_{m=1}^{\infty}$ of compact intervals such that for each m,

(i) $I_{m+1} \subset I_m$,

(ii) $|I_{m+1}| = \frac{1}{2}|I_m|$,

(iii) $\Sigma_{n=1}^{\infty}|I_m \cap J_n| < |I_m|$.

From (i) and (ii) we see that $\bigcap_{m=1}^{\infty}I_m$ contains a unique point x, which clearly belongs to I.

Now consider an arbitrary positive integer n and an arbitrary point y of J_n. Choose $r > 0$ so that $(y-r,y+r) \subset J_n$, and then choose m so that $|I_m| < r/2$. By (iii), $|I_m \cap J_n| < |I_m|$; whence, by (4.3), there exists z in $I_m - J_n$. Thus

$$|x - y| \geq |z - y| - |z - x| \geq r - |I_m| > r/2.$$

Since y is arbitrary, $x \in I - J_n$. Since n is arbitrary, $x \in \bigcap_{n=1}^{\infty}(I - J_n)$, and the proof is complete. □

Taken together, Theorems (4.1) and (4.5) provide a paradox which we should attempt to resolve. In doing this, first observe that

Theorem (4.1) is not really so surprising if we look at RUSS with a classical eye: the constructive real line is then seen to be countable, and so, naturally, has Lebesgue measure 0. However, these observations are false when the constructive real line is examined from within RUSS: for $\mathbb{R}$ is then seen to be uncountable, by Cantor's Theorem ((1.4) of Chapter 2), so there is no reason to expect that the constructive Lebesgue measure of the constructive real line will be 0.

So when does a subset S of $\mathbb{R}$ have constructive Lebesgue measure 0? It does so when for each $\epsilon > 0$ there is a sequence $(I_n)_{n=1}^{\infty}$ of bounded open intervals such that $S \subset \bigcup_{n=1}^{\infty} I_n$, and such that $\sum_{n=1}^{\infty}|I_n|$ converges to a sum less than ϵ. Theorem (4.5) shows, in a very positive manner, that [0,1] cannot have Lebesgue measure 0 in RUSS. Theorems (4.1) and (4.5) do not conflict, because the former does not produce a sequence of bounded open intervals (I_n) with $\sum_{n=1}^{\infty}|I_n|$ *convergent* and arbitrarily small. In fact, Theorem (4.5) shows that the series $\sum_{n=1}^{\infty}|I_n|$ in Theorem (4.1) cannot converge within RUSS.

Of course, in any sensible measure theory the interval [0,1] will have Lebesgue measure equal to 1. However there are a few surprises even in a sensible constructive measure theory. One such surprise is that compact subsets of $\mathbb{R}$ need not be measurable. It is fairly easy to construct a Brouwerian example of a nonmeasurable compact set; this example can be elaborated upon, to provide an example within RUSS of compact set that cannot have an outer measure.

Let $(r_n)_{n=0}^{\infty}$ be the enumeration of the rational points in [0,1], and let $(a_n)_{n=0}^{\infty}$ be a decreasing binary sequence. Then it is easy to check that the set $A = \{a_n r_n : n = 0,1,\ldots\}$ is a totally bounded subset of [0,1], so that its closure S is compact. (Classically, S is either finite or all of [0,1].) If S is measurable, then either $\mu(S) > 0$, in which case $a_n = 0$ for all n; or $\mu(S) < 1$, in which case the terms a_n cannot all be zero. Thus the proposition

Every compact subset of $\mathbb{R}$ *is Lebesgue measurable*

entails the omniscience principle

For each binary sequence (a_n), *either* $\forall n(a_n = 0)$ *or* $\neg\forall n(a_n = 0)$.

Within RUSS, we can modify this example to get a set whose measurability would be contradictory; viewed classically, this example will provide a set whose measure is not computable in the sense of recursive function theory.

The **outer measure** of a subset S of $\mathbb{R}$ is defined to be

$$\mu^*(S) \equiv \inf\{\Sigma_{n=1}^{\infty}(b_n - a_n) : S \subset \cup_{n=1}^{\infty}[a_n,b_n]\}$$

if the infimum exists. Finite sets clearly have outer measure 0, and, by (4.5), a compact interval $[a,b]$ has outer measure $b - a$. We will need the readily verifiable fact that if $\mu^*(S)$ exists, and I is a compact interval whose endpoints are in the metric complement of S, then $\mu^*(S \cap I)$ exists.

Now let

$$K \equiv \{n \in \mathbb{N} : \varphi_n(n) \text{ is defined}\}.$$

Then K is countable (Problem 6); let k_0, k_1, ... be an enumeration of K. Define a sequence $(J_n)_{n=0}^{\infty}$ of pairwise disjoint closed intervals by setting

$$J_n \equiv [2^{-n}, 2^{-n} + 3^{-n}]$$

for each n, and let $(r_n)_{n=0}^{\infty}$ be a dense sequence of rational numbers in $\cup_{n=0}^{\infty}J_n$. For each m in $\mathbb{N}$, there exists a unique n such that $r_m \in J_n$; if $n \in \{k_0,\ldots,k_m\}$, set $\theta_m \equiv 0$; otherwise, set $\theta_m \equiv r_m$. Let

$$A \equiv \{\theta_m : m \in \mathbb{N}\},$$

and let S be the closure of A in $\mathbb{R}$. Then S is clearly complete; the reader may verify that S is totally bounded, and hence compact. Suppose that $\mu^*(S)$ exists. Define a sequence $(q_n)_{n=0}^{\infty}$ of rational numbers by

$$q_n \equiv \mu^*(S \cap J_n).$$

Clearly, $q_n = 3^{-n}$ if $\varphi_n(n)$ is undefined, and $q_n = 0$ if $\varphi_n(n)$ is defined. Moreover, if $q_n > 0$, then $\varphi_n(n)$ cannot be defined; so $\varphi_n(n)$ is undefined, and therefore $q_n = 3^{-n}$. Thus if $q_n < 3^{-n}$, we cannot have $q_n > 0$, so that $q_n = 0$. It follows that $q_n \in \{0,3^{-n}\}$. Also, $q_n = 0$ if and only if $\varphi_n(n)$ is not undefined: for if $q_n = 0$, then $q_n \neq 3^{-n}$; while if $\varphi_n(n)$ is not undefined, then q_n cannot equal 3^{-n}, and so $q_n = 0$. Setting $g(n) \equiv 1 - 3^n q_n$, we obtain a contradiction to Lemma (1.2). This completes the proof that $\mu^*(S)$ cannot exist.

5. Moduli of continuity and cozero sets

Ceitin's theorem says not only that every function $f:\mathbb{R} \to \mathbb{R}$ is pointwise continuous, but that for each $f:\mathbb{R} \to \mathbb{R}$ we can construct a function that serves as a modulus of pointwise continuity for f. In this section we formulate some basic facts about such moduli, and make some observations about finite regular sequences.

Let $f:X \to Y$ be a map between metric spaces, and ϵ a positive number. A function $M:X \to \mathbb{R}^+$ is a **modulus of ϵ-continuity** for f if

$$\rho(x,y) < M(x) \Rightarrow \rho(f(x),f(y)) < \epsilon.$$

A function $M:X \times \mathbb{R}^+ \to \mathbb{R}^+$ is a **modulus of pointwise continuity** for f if $M(x,\epsilon)$, as a function of x, is a modulus of ϵ-continuity for f for each $\epsilon > 0$.

(5.1) Lemma *If $f:X \to Y$ has a modulus of ϵ-continuity for arbitrarily small $\epsilon > 0$, then f has a modulus of pointwise continuity.*

Proof. Let $(\epsilon_i)_{i=0}^{\infty}$ be a sequence of positive numbers converging to 0, such that $\epsilon_{i+1} \leq \epsilon_i/2$ for each i, and for each i let M_i be a modulus of ϵ_i-continuity for f. We may assume that $M_{i+1} \leq M_i \leq \epsilon_i$ for each i. Let

$$S \equiv \bigcup_{i=0}^{\infty}[\epsilon_{i+1},\epsilon_i] \cup [\epsilon_0,\infty),$$

and define $M(x,\epsilon)$ on $X \times S$ by setting $M(x,\epsilon_i) \equiv M_{i+2}(x)$, extending linearly over each interval $[\epsilon_{i+1},\epsilon_i]$, and setting $M(x,\epsilon) \equiv M_2(x)$ if $\epsilon \geq \epsilon_0$. As

$$M_{i+2}(x) - M_{i+3}(x) \leq M_{i+2}(x) \leq \epsilon_{i+2} < \epsilon_i - \epsilon_{i+1},$$

we have

$$|M(x,\epsilon) - M(x,\epsilon')| \leq |\epsilon - \epsilon'|$$

for all ϵ,ϵ' in S. Therefore $M(x,\epsilon)$ extends uniquely to a function on $X \times \mathbb{R}^+$ which is easily seen to be positive and increasing in ϵ. If $\epsilon > 0$, then there exists $i \geq 0$ such that $\epsilon_{i+1} < \epsilon < \epsilon_{i-1}$, where $\epsilon_{-1} \equiv \infty$. Then $M(x,\epsilon) \leq M_{i+1}(x)$; so if $\rho(x,y) < M(x,\epsilon)$, then $\rho(f(x),f(y)) < \epsilon_{i+1} < \epsilon$. □

We aim to develop a sufficient condition for the existence of a modulus of ϵ-continuity. To this end, we first define a finite sequence $(r_1,\ldots,r_n)$ of points in a metric space X to be **regular** if $\rho(r_i,r_j) \leq i^{-1} + j^{-1}$ for all $i,j \leq n$. If $(r_1,\ldots,r_n)$ is a finite regular sequence of points in X, then the finite intersection of closed balls

$$\bigcap_{i=1}^{n} \overline{B}(r_i, i^{-1} + (n+1)^{-1})$$

consists of exactly those $x \in X$ such that $(r_1,\ldots,r_n,x)$ is regular. We shall be interested in slightly more general sets than finite intersections of open balls.

A subset S of a metric space (X,ρ) is a **cozero set** if there is a pointwise continuous function $f:X \to \mathbb{R}$ such that $S = \{x \in X : f(x) \neq 0\}$. If, in addition, f is a **Lipschitz function**, in the sense that $|f(x) - f(y)| \leq \rho(x,y)$ for all x,y in X, then we say that S is a **Lipschitz cozero set**; in that case, replacing f by $\min\{1,|f|\}$ if necessary, we may assume that f is nonnegative and bounded by 1.

Classically, every open set in a metric space is a Lipschitz cozero set. However, an open set in an arbitrary topological space need not be a cozero set.

The simple proof of the following lemma is left to the reader.

(5.2) Lemma *If (a_n) is a bounded sequence of nonnegative numbers, then $\sup\{a_n/n : n \in \mathbb{N}^+\}$ exists.* □

(5.3) Proposition *Cozero sets are open. Every open ball in a metric space is a Lipschitz cozero set. Finite intersections, and countable unions, of Lipschitz cozero sets are Lipschitz cozero sets.*

Proof. We only prove the last statement, leaving the others as exercises. Let (f_n) be a sequence of pointwise continuous Lipschitz functions from X into $\mathbb{R}$, and for each n let $C_n \equiv \{x : f_n(x) \neq 0\}$. We may assume that each f_n is nonnegative and bounded by 1. Then

$$C_1 \cap C_2 \cap \ldots \cap C_n = \{x : \min\{f_1(x),\ldots,f_n(x)\} \neq 0\}$$

and

$$\bigcup_{n=1}^{\infty} C_n = \{x : \sup\{f_n(x)/n : n \in \mathbb{N}^+\} \neq 0\}$$

are clearly Lipschitz cozero sets. □

It follows from Proposition (5.3) that the Lipschitz cozero sets in $\mathbb{N}$ are precisely the countable subsets of $\mathbb{N}$.

We will apply the following lemma in the case where $X \equiv \mathbb{R}$, $Q \equiv \mathbb{Q}$, and the Lipschitz cozero sets are open intervals.

(5.4) Lemma *Let X be a metric space, Q a dense subset of X, f a function from X to a metric space, and $\epsilon > 0$. Suppose there exists a countable set $\mathcal{C}$ of Lipschitz cozero subsets of X such that*

(i) *for each $x \in X$, there exists $C \in \mathcal{C}$ such that $x \in C$, and such that if $q \in Q \cap C$, then $\rho(f(x),f(q)) < \epsilon/4$;*

(ii) *if $C \in \mathcal{C}$, and $q_1,q_2 \in Q \cap C$, then $\rho(f(q_1),f(q_2)) < \epsilon/2$.*

Then f has a modulus of ϵ-continuity.

Proof. Let D be a detachable subset of $\mathbb{N}$ that maps onto $\mathcal{C}$. For each i in D let C_i denote the image of i in $\mathcal{C}$, and let $\lambda_i : X \to \mathbb{R}$ be a nonnegative, pointwise continuous, Lipschitz function bounded by 1, such that $C_i = \{x \in X : \lambda_i(x) \neq 0\}$. Then

$$M(x) \equiv \sup\{\lambda_i(x)/i : i \in D\}$$

exists, by (5.2). For each x in X there exists i in D such that $x \in C_i$; whence $\lambda_i(x) > 0$, and therefore $M(x) > 0$. Consider x,y such that $\rho(x,y) < M(x)$. Choosing j in D so that $\rho(x,y) < \lambda_j(x)/j$, we have $x \in C_j$; also, as λ is a Lipschitz function,

$$\lambda_j(y) \geq \lambda_j(x) - \rho(x,y) > 0,$$

and so $y \in C_j$. By (i), there exist $m,n \in D$ such that $x \in C_m$ and $y \in C_n$, and such that if $q_1 \in Q \cap C_m$ and $q_2 \in Q \cap C_n$, then $\rho(f(x),f(q_1)) < \epsilon/4$ and $\rho(f(y),f(q_2)) < \epsilon/4$. As Q is dense in X, and cozero sets are open, we can find $q_1 \in Q \cap C_j \cap C_m$ and $q_2 \in Q \cap C_j \cap C_n$. Then by (ii),

$$\begin{aligned}\rho(f(x),f(y)) &\leq \rho(f(x),f(q_1)) + \rho(f(q_1),f(q_2)) + \rho(f(q_2),f(y)) \\ &< \epsilon/4 + \epsilon/2 + \epsilon/4 = \epsilon,\end{aligned}$$

so that M is a modulus of ϵ-continuity for f. □

6. Ceitin's theorem

We need more than CPF to prove Ceitin's theorem that every function from $\mathbb{R}$ to $\mathbb{R}$ is continuous. One essential ingredient is Markov's principle, which we have already discussed. The other concerns the Russian constructivists' belief that everything is a natural number. To reflect that belief within our framework, we must either restrict Ceitin's theorem to a certain class of functions from $\mathbb{R}$ to $\mathbb{R}$, or postulate that every function is in that class. Although the latter is more in the spirit of the philosophy of RUSS, we shall adopt the former, more conservative, point of view.

Markov's principle enters through a technique that Ceitin called the *capture method*.

(6.1) Lemma *Let A and B be countable subsets of* $\mathbb{N}$ *such that the complement of A is contained in B and is not countable. If Markov's* *principle holds, then* $A \cap B$ *is nonvoid.*

Proof. The intersection of the countable sets A and B is countable. Let D be a detachable subset of $\mathbb{N}$ that is mapped onto $A \cap B$. If D were empty, then B would be the complement of A, which is not countable. Thus D is nonempty, by Markov's principle. □

We now prove a simple lemma about countable subsets of $\mathbb{N}^r$.

(6.2) Lemma *Let* $r,s \in \mathbb{N}^+$, *let* A *be a subset of* $\mathbb{N}^r$, *and let* σ *be a mapping of* $\mathbb{N}^r \times \mathbb{N}^s$ *into* $\{0,1\}$ *such that* $x \in A$ *if and only if there exists* $y \in \mathbb{N}^s$ *with* $\sigma(x,y) = 1$. *Then* A *is countable.*

Proof. The set $S \equiv \{(x,y) \in \mathbb{N}^r \times \mathbb{N}^s : \sigma(x,y) = 1\}$ is a detachable subset of $\mathbb{N}^r \times \mathbb{N}^s$, and so is countable. As A is the image of S under the projection onto the first coordinate, the result follows. □

For the work of this section, it is convenient to consider the functions φ_n as being defined on $\mathbb{N}^+$ rather than $\mathbb{N}$.

A function $f:\mathbb{R} \to \mathbb{R}$ is said to be **representable** if there exists a partial function $F:\mathbb{N} \to \mathbb{N}$ with countable domain, such that if $\varphi_n:\mathbb{N} \to \mathbb{Q}$ is a regular Cauchy sequence converging to $r \in \mathbb{R}$, then $\varphi_{F(n)}:\mathbb{N} \to \mathbb{Q}$ is a regular Cauchy sequence converging to $f(r)$. Note that $F(n)$ may be defined even if φ_n is not regular, or even total.

In the standard presentations of RUSS, every function is representable, *by definition*. Even without this assumption, every concrete function that we construct within RUSS turns out to be representable. However we will not make the sweeping assumption that every function is representable, as our only application of representability is to Ceitin's theorem. Instead, we will state Ceitin's theorem in terms of representable functions.

In the following, υ is a mapping of $\mathbb{Q}$ into $\mathbb{N}$ such that for each q in $\mathbb{Q}$, the function $\varphi_{\upsilon(q)}$ is identically equal to q.

(6.3) Lemma *Let* $F:\mathbb{N} \to \mathbb{N}$ *be a partial function with countable domain, and for all* $N,p,m \in \mathbb{N}$ *let*

$$\begin{aligned} L(N,p,m) = \{q \in \mathbb{Q} : \; & p \in \operatorname{dom} F \text{ and } N \in \operatorname{dom} \varphi_{F(p)}, \\ & |\varphi_{F(\upsilon q)}(N) - \varphi_{F(p)}(N)| > 4/N, \text{ and} \\ & (\varphi_p(1),\ldots,\varphi_p(m),q) \text{ is defined and regular}\}. \end{aligned}$$

Then the sets $\{(N,p,m,q) : q \in L(N,p,m)\}$ *and* $E \equiv \{q : q \in L(N,p,m)$ *for some* $N,p,$ *and* $m\}$ *are countable.*

Proof. Write $F = \varphi_t$. Define $\sigma(N,p,m,q,k) \equiv 1$ if $p \in D_t(k)$, $N \in D_{F(p)}(k)$, $|\varphi_{F(vq)}(N) - \varphi_{F(p)}(N)| > 4/N$, $i \in D_p(k)$ for each i in $\{1,\dots,m\}$, and $(\varphi_p(1),\dots,\varphi_p(m),q)$ is regular; and define $\sigma(N,p,m,q,k) \equiv 0$ otherwise. Then $q \in L(N,p,m)$ if and only if there exists k such that $\sigma(N,p,m,q,k) = 1$. The result now follows from (6.2). □

(6.4) Corollary *Under the conditions of Lemma (6.3), there exists a partial function* $\ell:\mathbb{N}^3 \to \mathbb{Q}$ *with countable domain, such that* $\ell(N,p,m) \in L(N,p,m)$ *if* $L(N,p,m)$ *is nonvoid, and* $\ell(N,p,m)$ *is undefined otherwise.*

Proof. By (6.3), for all N,p,m in $\mathbb{N}$ there exist a detachable subset $D(N,p,m)$ of $\mathbb{N}$, and a mapping $e(N,p,m)$ of $D(N,p,m)$ onto

$$E(N,p,m) \equiv \{q : q \in L(N,p,m)\}.$$

Define $\sigma:\mathbb{N}^3 \times \mathbb{N} \to \{0,1\}$ by

$$\begin{aligned}\sigma(N,p,m,k) &= 1 \quad \text{if } k \text{ is the least element of } D(N,p,m),\\ &= 0 \quad \text{otherwise.}\end{aligned}$$

Then by (6.2),

$$A \equiv \{(N,p,m) : \sigma(N,p,m,k) = 1 \text{ for some } k\}$$

is countable. For each (N,p,m) in A define $\ell(N,p,m) \equiv e(k)$, where k is the least element of $D(N,p,m)$. □

We now use the sets $L(N,p,m)$ of Lemma (6.3), and the function $\ell(N,p,m)$ of Corollary (6.4), to establish **Ceitin's theorem:**

(6.5) Theorem *If Markov's principle holds, then every representable function from* $\mathbb{R}$ *to* $\mathbb{R}$ *is pointwise continuous.*

Proof. Given a representable function $f:\mathbb{R} \to \mathbb{R}$ and a positive integer N, we shall construct a countable set $\mathcal{C}$ of open intervals in $\mathbb{R}$ with rational endpoints, such that the hypotheses of (5.4) are satisfied with $X \equiv \mathbb{R}$, $Q \equiv \mathbb{Q}$, and $\epsilon > 24/N$. The theorem will then follow from (5.4) and (5.1).

Let F represent f, and define a partial function θ from $\mathbb{N}^3$ to $\mathbb{Q}$ by

$$\begin{aligned}\theta(n,p,s) &= \varphi_p(s) &&\text{if } n \notin D_n(s-1),\\ &= \ell(N,p,m) &&\text{if } n \in D_n(m)\backslash D_n(m-1) \text{ and } s > m,\end{aligned}$$

where $\ell(N,p,m)$ is as in (6.4). Using the s-m-n theorem (1.7), construct a total function $\gamma : \mathbb{N}^2 \to \mathbb{N}$ such that $\theta(n,p,s) = \varphi_{\gamma(n,p)}(s)$. Define

$$S_p \equiv \{n : p \in \operatorname{dom} F,\ \gamma(n,p) \in \operatorname{dom} F,\ N \in \operatorname{dom} \varphi_{F(p)} \cap \operatorname{dom} \varphi_{F\gamma(n,p)}, \text{ and } |\varphi_{F\gamma(n,p)}(N) - \varphi_{F(p)}(N)| < 2/N\}$$

and

$$C_p \equiv \{m : n \in D_n(m) \backslash D_n(m-1) \text{ for some } n \in S_p, \text{ and } (\varphi_p(1),\ldots,\varphi_p(m)) \text{ is defined and regular}\}$$

Simple applications of (6.2) show that S_p and C_p are countable. For each $m \in C_p$, let

$$a \equiv \max\{\varphi_p(i) - i^{-1} - (m+1)^{-1} : i = 1,\ldots,m\},$$
$$b \equiv \min\{\varphi_p(i) + i^{-1} + (m+1)^{-1} : i = 1,\ldots,m\},$$
$$\text{and } I_{p,m} \equiv (a,b).$$

Note that $\mathbb{Q} \cap I_{p,m} \subseteq \{q \in \mathbb{Q} : (\varphi_p(1),\ldots,\varphi_p(m),q) \text{ is regular}\}$, and that if φ_p is a regular Cauchy sequence, then its limit belongs to $I_{p,m}$ for each m.

Finally, let

$$\mathscr{C} \equiv \{I_{p,m} : p \in \mathbb{N},\ m \in C_p\}.$$

The rest of the proof consists of several steps.

(6.5.1) *If $n \in D_n(m) \backslash D_n(m-1)$ for some $n \in S_p$, then $L(N,p,m)$ is empty.*

For if $L(N,p,m)$ is nonempty, $\varphi_{\gamma(n,p)}$ is a regular Cauchy sequence with all but finitely many terms equal to $q \equiv \ell(N,p,m)$; as $q \in L(N,p,m)$, we then have $p \in \operatorname{dom} F$, $N \in \operatorname{dom} \varphi_{F(p)}$, and $|\varphi_{F(\upsilon q)}(N) - \varphi_{F(p)}(N)| > 4/N$. But $|\varphi_{F\gamma(n,p)}(N) - \varphi_{F(p)}(N)| < 2/N$, because $n \in S_p$. Therefore

$$|\varphi_{F\gamma(n,p)}(N) - \varphi_{F(\upsilon q)}(N)| > 4/N - 2/N = 2/N.$$

This is impossible, as $\varphi_{\gamma(n,p)}$ and $\varphi_{\upsilon q}$ are regular Cauchy sequences converging to q, and therefore $\varphi_{F\gamma(n,p)}$ and $\varphi_{F(\upsilon q)}$ are regular Cauchy sequences converging to $f(q)$. Thus $L(N,p,m)$ is empty.

(6.5.2) *If $m \in C_p$ and $q \in \mathbb{Q} \cap I_{p,m}$, then $|\varphi_{F(\upsilon q)}(N) - \varphi_{F(p)}(N)| \leq 4/N$.*

For if $m \in C_p$, then $n \in D_n(m) \backslash D_n(m-1)$ for some $n \in S_p$. Therefore S_p is nonempty, so that $p \in \operatorname{dom} F$ and $N \in \operatorname{dom} \varphi_{F(p)}$. If also $q \in \mathbb{Q} \cap I_{p,m}$, then $(\varphi_p(1),\ldots,\varphi_p(m),q)$ is defined and regular. The desired conclusion now follows from the definition of $L(N,p,m)$, since that set is empty by

(6.5.1).

In view of (5.1), it only remains to prove

(6.5.3) *f has a modulus of ϵ-continuity.*

If $q_1, q_2 \in \mathbb{Q} \cap I_{p,m}$, then it follows from (6.5.2) that

$$|\varphi_{F(\nu q_1)}(N) - \varphi_{F(\nu q_2)}(N)| \leq 8/N,$$

and therefore $|f(q_1) - f(q_2)| \leq 10/N < \epsilon/2$. This verifies (5.4,ii). On the other hand, suppose φ_p is a regular Cauchy sequence, and write

$$r \equiv \lim_{n\to\infty} \varphi_p(n).$$

We shall show that

$$S_p \supset K' \equiv \{n : \varphi_n(n) \text{ is not defined}\}.$$

If $n \in K'$, then $\varphi_{\gamma(n,p)} = \varphi_p$ is a regular Cauchy sequence, by definition of γ; whence $\gamma(n,p) \in \operatorname{dom} F$, $\varphi_{F(\gamma(n,p))}$ and $\varphi_{F(p)}$ are regular Cauchy sequences converging to $f(r)$, and so $|\varphi_{F(\gamma(n,p))}(s) - \varphi_{F(p)}(s)| \leq 2/s$ for all s. Now, K' is not countable (why?). So, applying (6.1) with $A \equiv \{n : \varphi_n(n) \text{ is defined}\}$ and $B \equiv S_p$, we see that $A \cap S_p$ is nonvoid; whence there exists $m \in C_p$. Then $r \in I_{p,m}$. So if $q \in \mathbb{Q} \cap I_{p,m}$, then (6.5.2) gives

$$\begin{aligned} |f(q) - f(r)| &\leq |f(q) - \varphi_{F(\nu q)}(N)| + |\varphi_{F(\nu q)}(N) - \varphi_{F(p)}(N)| \\ &\qquad + |\varphi_{F(p)}(N) - f(r)| \\ &< 1/N + |\varphi_{F(\nu q)}(N) - \varphi_{F(p)}(N)| + 1/N \\ &< 6/N < \epsilon/4. \end{aligned}$$

This verifies (i) of Lemma (5.4). Reference to that lemma now completes the proof of Ceitin's theorem. □

PROBLEMS

1. Prove that the set of partial function algorithms is not countable.
2. Prove that Markov's principle is equivalent to the following statement: For all partial functions $f : \mathbb{N} \to \mathbb{N}$ and all n in $\mathbb{N}$, if $f(n)$ is not undefined, then $f(n)$ is defined.
3. Prove that each of the following statements is equivalent to

Markov's principle.

(i) For each real number x, $\neg(x \geq 0) \Rightarrow (x < 0)$.

(ii) For each sequence (a_n) of real numbers and each b in $\mathbb{R}$, if $a_n \geq b$ for all n, then for each $\epsilon > 0$ there exists n such that $a_n > b - \epsilon$.

4. Show, within BISH, that if there is a countable subset K of $\mathbb{N}$ such that $\mathbb{N}\backslash K$ cannot be countable, then the capture method (6.1) implies Markov's principle. (Hint: Let α be a binary sequence that cannot be all zeros, and let $A \equiv \{x \in K : \alpha_n = 1$ for some $n\}$. Show that A is countable and that $\mathbb{N}\backslash A = \mathbb{N}\backslash K$. Apply the capture method to A and $\mathbb{N}$.)

5. Prove the **recursion theorem:** if $f:\mathbb{N} \to \mathbb{N}$ is a total function, then there exists n such that $\varphi_n = \varphi_{f(n)}$.

6. Let $K \equiv \{n \in \mathbb{N} : \varphi_n(n)$ is defined$\}$. Show that K is countable. Show also that the complement of K is **productive**, in the following sense: if (n_k) is a sequence of nonnegative integers such that $n_k \notin K$ for all k, then there exists N in $\mathbb{N}$ such that (i) $N \notin K$, and (ii) $N \neq n_k$ for all k. Hence show that K is not detachable from $\mathbb{N}$.

7. Show that $\{n \in \mathbb{N} : \varphi_n(n)$ is not defined$\}$ is not countable.

8. Let A be a nonempty subset of $\mathbb{N}$, and suppose that there exists a mapping f of A onto the power set $\mathcal{P}(\{1\})$ of $\{1\}$. Prove that for each subset B of $\mathbb{N}$, there exists a map $g:\mathbb{N} \to \mathbb{N}$ such that for all n in $\mathbb{N}$, $n \in B$ if and only if $f(g(n)) = \{1\}$. Taking

$$B \equiv \{n \in \mathbb{N} : \varphi_n(n) \text{ is defined and } \neg(f(\varphi_n(n)) = \{1\})\},$$

deduce a contradiction. (Thus we obtain a result due to Myhill: the power set of a singleton is not denumerable.)

9. Construct examples of each of the following:

(i) a pointwise continuous map $f:[0,1] \to [0,1]$, with supremum 1, such that $f(x) < 1$ for each x;

(ii) a pointwise continuous map of $[0,1]$ into $\mathbb{R}$ that assumes every positive integral value;

(iii) a pointwise continuous map of $[0,1]$ into itself that does not have a supremum.

10. The interval $[0,1]$ is not connected: Construct a nonvoid subset S of $[0,1]$ such that (i) S is both open and closed in $[0,1]$, and (ii) $[1/2,1] \subset [0,1]-S$.

11. Let (r_n) be the sequence of Specker's theorem, and K the nonvoid closed subset $\{x \in \mathbb{R} : x \leq r_n \text{ for some } n\}$ of $[0,1]$. Show that the following statements are equivalent.

    (i) For each x in $\mathbb{R}$ there exists y in K such that if $x \neq y$, then $x \neq z$ for all z in K.

    (ii) K is totally bounded.

    Deduce that (i) is false. (*cf*. Lemma (3.3) of Chapter 2.)

12. Construct a sequence (K_n) of nonempty closed subsets of $[0,1]$ such that (i) $K_1 \supset K_2 \supset K_3 \supset \ldots$, and (ii) $\cap_{n=1}^{\infty} K_n$ is empty.

13. Construct a sequence (f_n) of uniformly continuous mappings from $[0,1]$ to itself such that (i) (f_n) converges pointwise to 0 on $[0,1]$, and (ii) $\sup f_n = 1$ for each n.

NOTES

The general approach of this chapter is based on Richman's paper *Church's thesis without tears* (J. Symbolic Logic 48 (1983), 797–803). The axiom CPF is an adaptation of the abstract computer science idea of an **acceptable programming system:** see M. Machtey and P. Young, *An introduction to the general theory of algorithms* (North-Holland, 1978). The name CPF is an acronym from Countable Partial Functions.

We have benefited greatly from reading O. Aberth's *Computable analysis* (McGraw-Hill, 1980); other useful references for RUSS are **Kushner** and *Constructive real numbers and constructive function spaces*, by N.A. Shanin (AMS Translations 21, 1968). The interested reader is encouraged to consult the innumerable papers of O. Demuth on analysis within the framework of RUSS.

Corollary (1.6) is Theorem 4, page 142, in **Kushner**. Kushner's proof is based on a more positive result, found in Ceitin, *Mean value theorems in constructive analysis* (published in Russian in 1962, and in AMS Translations 64 in 1967).

Our notion of an *operation*, defined in Section 2, is a

substitute for Bishop's notion of an operation as a finite procedure which assigns to each element of a set A an element of a set B, but which need not assign equal elements to equal elements.

Theorem (2.3) is due to A. A. Markov, *On the continuity of constructive functions* (in Russian, in Uspekhi Mat. Nauk 9 (1954), 226–230); it is Theorem 7.1 in Aberth. Theorem (2.4) is folklore among constructive mathematicians, who often use the intermediate value theorem as an example of a classical theorem that is not constructively valid.

Specker's theorem appears in E. Specker, *Nicht konstruktiv beweisbare Sätze der Analysis* (J. Symbolic Logic 14 (1949), 145–158).

Sensible constructive measure theories are found in Chapter 6 of **Bishop–Bridges**; Chapter 7 of **Kushner**; Chapter VI of Heyting's book *Intuitionism – an Introduction* (3rd. edn., North-Holland, 1971); and the paper *Remarks on constructive mathematical analysis*, by O.Demuth and A. Kucera, in *Logic Colloquium* '78 (M. Boffa, D van Dalen, and K. McAloon, eds, North-Holland, 1979).

The proof of Ceitin's theorem in Section 6 is a barely recognizable adaptation of the proof in the third appendix of Aberth's book.

Chapter 4. Constructive Algebra

In which various examples are given to illustrate the problems and techniques of constructive algebra. As in Chapter 2, particular attention is paid to results whose classical counterparts are either routine or trivial, but whose constructive proof requires careful choice of definitions and hypotheses, and subtle technique. For example, in Section 6 there is given a constructive version of the Hilbert basis theorem, for which we need both the correct definition of a Noetherian ring, and the classically redundant hypothesis of coherence. The relationship between constructive and recursive algebra is illustrated in Section 4, which deals with the question of the uniqueness of splitting fields.

1. General considerations

Although several mathematicians, including Kronecker and van der Waerden, have made important contributions to constructive algebra, the main thrust of constructive mathematics has been in the direction of analysis. One reason for this is that all analysis is based on the real numbers, where constructive difficulties arise at the very outset; vast areas of algebra, on the other hand, deal with finite objects, such as polynomials over $\mathbb{Q}$, and even with finite sets of objects, as in the theory of finite groups. Another possible explanation is that the main research interests of Brouwer, Markov, and Bishop lay outside algebra.

In 1941, Heyting published *Untersuchungen über intuitionistische Algebra*, a paper concerned with vector spaces and polynomial rings over nondiscrete fields such as $\mathbb{R}$. Although the theorems in that paper specialize to discrete fields, the interest is in the nondiscrete case; much effort and ingenuity is spent on the inequality relation, and on other considerations that arise because of nondiscreteness. The resulting axiomatic treatment has the flavour of rings of continuous functions, or of finite-dimensional Banach spaces, rather than of purely algebraic constructions such as discrete abelian groups, or finite-dimensional algebras over discrete fields.

Perhaps it is more accurate to say that *discrete* algebra has been neglected by constructive mathematicians, who found the nondiscrete problems more challenging. In contrast, discrete algebra received much attention from algorithmically oriented classical mathematicians in the context of **recursive mathematics**, which we may think of as RUSS, with classical logic, interpreted as referring to *computable* functions, rather than to all functions. We will observe the differences between recursive and constructive algebra as we proceed.

We assume that the reader is familiar with basic algebraic structures such as rings, ideals, and modules; these structures involve only functions and equations. When we come to the subject of fields, we are interested in invertible elements, that is, elements x such that $xy = yx = 1$ for some y. Although invertibility is a purely algebraic notion, fields require the property

(*) if $x \neq 0$, then x is invertible;

thus in this context we must talk about an inequality as well as an equality. If we suppose that the underlying set is discrete, then (*) poses no problem, and by appending it to the definition of a commutative ring with 1, we get the definition of a **discrete field**. The same issue arises with the notion of an *integral domain*.

In this chapter, fields and integral domains, but *not* rings or modules, will be assumed discrete; and all modules will be left modules. We shall be interested mainly in commutative rings, but where we can include the noncommutative case for free, we shall do so.

2. Factoring

We consider the problem of factoring in discrete integral domains; in particular, we will be interested in the ring $\mathbb{Z}$ of integers, and the ring $k[X]$ of polynomials in one variable over a discrete field k.

First, we need some standard terminology. An element r of a discrete integral domain R is **invertible**, or a **unit**, if there exists $s \in R$, called the **inverse** of r, such that $rs = 1$. Two elements r and s of R are **associates**, which we write $r \sim s$, if $r = us$ for some unit u. A noninvertible element r is **irreducible** if whenever $r = ab$, then either a or b is invertible - that is, r has no **proper factors**. A discrete integral domain R is a **unique factorization domain** if

(i) each nonzero element of R is either invertible, irreducible, or a product of irreducibles; and

(ii) if $a_1 a_2 \cdots a_m = b_1 b_2 \cdots b_n$ for irreducible a_i and b_j, then $m = n$ and, after reindexing if necessary, $a_i \sim b_i$ for each i.

Both $\mathbb{Z}$ and $k[X]$ satisfy condition (ii). The proof of this relies on the Euclidean algorithm, and is pretty much the same in BISH as it is in classical mathematics. Condition (i) is much more of a problem in BISH; indeed, we will construct a Brouwerian example of a discrete field k such that (i) does not hold for $k[X]$. However, following Kronecker, we can verify (i) for $k[X]$ when k is the rational number field $\mathbb{Q}$ (Theorem 2.2).

From a classical point of view, it is as easy to verify (i) for $\mathbb{Q}[X]$ as for $\mathbb{Z}$. In each ring we have an integral measure of size - the degree of a polynomial in the first case, and the absolute value of an integer in the second - such that the size of a nonzero element is greater than the size of any proper factor. A given nonzero element is either invertible, or irreducible, or a product of proper factors. By induction on size, every element is of the desired form.

More is required from a constructive point of view: given a nonzero element, we must either find an inverse for it, or show that it is irreducible, or find a proper factor of it. This is easy, in principle, for the ring of integers: the only invertible integers are 1 and -1, and by attempting to divide a given integer $r > 1$ by each positive integer d such that $d^2 \leq r$, we either show that r is irreducible, or find a proper factor of r. In the ring $\mathbb{Q}[X]$ the invertible elements are the polynomials of degree 0, which are readily recognized, but the number of potential factors of a given polynomial is infinite, *a priori*, so we cannot test each candidate to see if it is a proper factor, as we did for the integers.

We call a discrete field k **factorial** if each polynomial of positive degree in $k[X]$ is either irreducible or (uniquely) a product of irreducible polynomials. As (ii) always holds for $k[X]$, the field k is factorial if and only if $k[X]$ is a unique factorization domain.

Before examining Kronecker's proof that $\mathbb{Q}$ is factorial, we construct a Brouwerian example, similar to one given by van der Waerden, of a countable discrete field k that is not factorial.

(2.1) Theorem *If each countable subfield of the Gaussian number field* $\mathbb{Q}(i)$ *is factorial (where* $i^2 = -1$*), then* LPO *holds.*

Proof. Fix an enumeration $q_1, q_2, \ldots$ of $\mathbb{Q}$, and an enumeration $\xi_1, \xi_2, \ldots$ of $\mathbb{Q}(i)$. Given a binary sequence (a_n) with at most one term equal to 1, enumerate a subfield $k \equiv \{x_1, x_2, \ldots\}$ of $\mathbb{Q}(i)$ by setting

$$\begin{aligned} x_n &= q_n && \text{if } a_j = 0 \text{ for each } j \le n, \\ &= \xi_{n-j+1} && \text{if } j \le n \text{ and } a_j = 1. \end{aligned}$$

If the polynomial $X^2 + 1$ is irreducible in $k[X]$, then $i \notin k$, and so $a_n = 0$ for all n; if $X^2 + 1$ factors in $k[X]$, then $i \in k$, and so we can find n such that $a_n = 1$. □

We solve the factoring problem in $\mathbb{Q}[X]$ by solving it in $\mathbb{Z}[X]$, the polynomial ring with integer coefficients, and showing that any element in $\mathbb{Z}[X]$ that has a proper factor in $\mathbb{Q}[X]$ has a proper factor in $\mathbb{Z}[X]$. The transition from $\mathbb{Q}[X]$ to $\mathbb{Z}[X]$ relies on **Gauss's lemma**, which says that the product of two primitive polynomials is primitive. (Recall that a polynomial in $\mathbb{Z}[X]$ is **primitive** if its coefficients have no common noninvertible factor).

Do we need to find a constructive proof of Gauss's lemma? After all, it is classically true, and we can determine whether or not a given polynomial in $\mathbb{Z}[X]$ is primitive or not. In a more general setting, the question is whether, given a binary sequence and a classical proof that each element is 0, we can conclude that each element is 0. If some element of the sequence were 1, then classical mathematics would be inconsistent. Whether or not it is impossible for classical mathematics to be inconsistent from a constructive point of view, it is a peculiar consideration to bring in when all we are interested in is Gauss's lemma. The fact is that, in cases like these, many of the classical proofs are constructive as they stand, or can be trivially modified so as to become constructive proofs. So our answer to the question posed at the beginning of this paragraph is "Yes, but it's easy to do." (Problem 5).

We now return to **Kronecker's theorem:**

(2.2) Theorem *The field $\mathbb{Q}$ is factorial.*

Proof. Let f be a nonzero polynomial in $\mathbb{Q}[X]$ of degree n. If $n = 0$, then f is a nonzero constant, so f is invertible; if $n = 1$, then f is irreducible. It suffices to show that if $n > 1$, then f is either irreducible or is a product of polynomials of smaller degree. We may assume that f is a primitive polynomial in $\mathbb{Z}[X]$.

Consider the integers $f(0)$, $f(1)$, ..., $f(n)$. If $f(m) = 0$ for some m, then $f(X) = (X - m)g(X)$, so we may assume that $f(m) \neq 0$ for $m = 0,\ldots,n$. We shall construct a finite list of polynomials in $\mathbb{Z}[X]$ such that every factor of f in $\mathbb{Z}[X]$ is among them; by attempting to divide f by each element of this list, we will either write f as a product of polynomials of smaller degree, or show that f is irreducible in $\mathbb{Z}[X]$.

For each function α from $\{0,\ldots,n\}$ to $\mathbb{Z}$ such that $\alpha(m)$ is a factor of $f(m)$ for $m = 0,\ldots,n$, we construct the unique polynomial f_α in $\mathbb{Q}[X]$ of degree at most n such that $f_\alpha(m) = \alpha(m)$ for $m = 0,\ldots,n$. The set of such f_α that have integer coefficients is the desired list. Note that the set of functions f_α is finite because $f(m)$ has only finitely many factors for each m.

Finally we show that if f has a proper factor in $\mathbb{Q}[X]$, then f has a proper factor in $\mathbb{Z}[X]$. If $f = gh$, where g and h are in $\mathbb{Q}[X]$, then we can write $g = ag_0$ and $h = bh_0$, where g_0 and h_0 are primitive polynomials in $\mathbb{Z}[X]$, and $a,b \in \mathbb{Q}$. Thus $f = (ab)g_0h_0$. Since g_0h_0 is primitive by Gauss's lemma, $ab \in \mathbb{Z}$; so we have a factorisation of f in $\mathbb{Z}[X]$. □

3. Splitting fields

If k is a discrete field, and $f(X)$ an irreducible polynomial with coefficients in k, then we can construct a discrete field containing k in which $f(X)$ has a root. This construction, due to Kronecker, is effected by letting K be the polynomial ring $k[Y]$ modulo the principal ideal generated by $f(Y)$ – that is, two elements of K are considered equal if their difference is divisible by $f(Y)$. It is easy to verify that K is a discrete field containing k, and that Y, considered as an element of K, is a root of $f(X)$.

We can effect the preceding construction whether or not $f(X)$

is irreducible, but we can't show that K is a field unless we know that $f(X)$ is irreducible. To find an extension field in which an arbitrary polynomial $f(X)$ of degree greater than zero has a root, the classical procedure is to factor $f(X)$ into irreducible polynomials, and then construct an extension field in which one of those factors has a root. That option is open to us if we can factor $f(X)$, but in general we must rely on other techniques. If k is countable, we can consider $k[Y]$ modulo a *maximal ideal* containing $f(Y)$, rather than the principal ideal generated by an irreducible factor of $f(Y)$.

A detachable proper ideal M in a commutative ring is **maximal** if $1 \in I$ whenever I is an ideal containing M and $I \setminus M$ is nonempty. Equivalently, M is maximal if and only if for each $r \in R$, either $r \in M$ or else $rs-1 \in M$ for some $s \in R$ – that is, the quotient ring R/M is a discrete field. We will construct a maximal ideal of $k[X]$ as the union of a sequence of finitely generated ideals. During this construction we will need to know that finitely generated ideals of $k[X]$ are detachable. This follows from the fact, which we now establish, that finitely generated ideals of $k[X]$ are principal – that is, $k[X]$ is a **Bézout domain.**

(3.1) Lemma *If k is a discrete field, then finitely generated ideals of $k[X]$ are principal, and therefore detachable.*

Proof. Let I be the ideal generated $f_1,\ldots,f_n$. If $n = 1$, then I is principal by definition. Otherwise, the Euclidean algorithm constructs r and s in $k[X]$ such that $d = rf_1 + sf_2$ divides both f_1 and f_2. Then I is generated by $d,f_3,\ldots,f_n$, and we can complete the proof that I is principal by induction on n. Finally, a polynomial g is in the principal ideal generated by f if and only if g is divisible by f, which is decidable. □

(3.2) Theorem *If k is a countable discrete field, and $f(X)$ a polynomial in $k[X]$ of degree greater than zero, then there is a countable discrete field containing k in which $f(X)$ has a root.*

Proof. We construct a countable detachable ideal M of $k[Y]$ as follows. Let $g_1(Y)$, $g_2(Y)$, ... be an enumeration of $k[Y]$, let I_0 be the ideal generated by $f(Y)$, and note that $1 \notin I_0$, as f has positive degree. For each $n \geq 1$, let J_n be the ideal generated by I_{n-1} and $g_n(Y)$; as J_n is finitely generated, we can use (3.1) to decide whether or not $1 \in J_n$. Let

$I_n \equiv J_n$ if J_n is proper, and $I_n \equiv I_{n-1}$ otherwise. The union M of the ideals I_n is detachable, because we can refer to the nth construction step to determine whether or not $g_n(Y)$ is in M. By construction, M does not contain 1. If I is an ideal of $k[Y]$ containing M, and if $g_m(Y) \in I \setminus M$, then $g_m(Y) \notin I_m$, so that $1 \in J_m \subset I$. Thus M is maximal, so that $k[Y]/M$ is a discrete field. Clearly, $k[Y]/M$ is countable; and Y, considered as an element of $k[Y]/M$, is a root of $f(X)$ in $k[Y]/M$. □

The countability condition of Theorem 3.2 is essential: it can be shown that if we could eliminate it, then we could prove the following primitive choice axiom, affectionately called **the world's simplest axiom of choice:**

> *Let S be a set such that each member of S, if any, is a two-element set, and any two elements of S are equal; then there is a* ***choice function*** *on S - that is, a function f from S to the union of the sets in S, such that $f(x) \in x$ for each x in S.*

We shall see in Chapter 7 that there is topos model in which this axiom fails.

By repeated application of Theorem 3.2, we can construct a **splitting field** for $f(X)$ - that is, a discrete field K containing k that is generated over k by roots of $f(X)$, and over which $f(X)$ is a product of linear factors. By a more elaborate repetition, we can construct an **algebraic closure** of k - that is, an extension K of k, such that (i) every element of K satisfies a nonzero polynomial with coefficients in k, and (ii) every polynomial with coefficients in K is a product of linear factors. Classically, any two splitting fields of a polynomial in $k[X]$, and any two algebraic closures of k, are isomorphic over k. We shall see in the next section that, from the point of view of BISH, we cannot hope to construct such an isomorphism between two splitting fields of an arbitrary polynomial.

The situation is different in recursive algebra. For one thing, there is no problem about the *existence* of a splitting field, in fact a finite-dimensional one, for a given polynomial f of degree n in $k[X]$. Such a field K can be specified by n^3 elements $c_{ijk} \in k$ that tell us how to multiply two elements of $K = k^n$. Once we know the c_{ijk}, we can calculate virtually anything we want in K, including inverses. With n^2

more elements we can specify the roots of f in K. As these $n^3 + n^2$ elements exist classically, there is no problem getting your hands on them in recursive algebra. Similarly an isomorphism between two finite-dimensional splitting fields for f over k is specified by a finite matrix of elements of k, and since this matrix exists classically, it exists in the context of recursive algebra.

However, if we could prove in BISH that any two splitting fields of a polynomial over a countable discrete field k were isomorphic over k, then, using countable choice, we could prove that, given sequences of countable discrete fields k_n, polynomials p_n, and splitting fields E_n and F_n for p_n over k_n, we could construct a sequence of isomorphisms $\varphi_n : E_n \to F_n$ over k_n. Although the former theorem is trivial in recursive algebra, the latter is false: we shall show that it is false even in RUSS (Corollary (4.7) below).

There is also a problem in recursive algebra if we want to construct an isomorphism between two algebraic closures of k. In fact Metakides and Nerode have shown, in the context of recursive algebra, that if any two algebraic closures of k are isomorphic over k, then k is factorial.

4. Uniqueness of splitting fields

We have seen in the preceding section how to construct a countable discrete splitting field for a polynomial over a countable discrete field. In this section, we shall show that LLPO is equivalent to the *uniqueness* of splitting fields for polynomials over countable discrete fields. We shall also investigate the relationship between the constructive and the recursive treatments of splitting fields.

An ideal P in a discrete ring is **prime** if $xy \in P$ entails either $x \in P$ or $y \in P$.

If k is a discrete field, and M a nonzero proper ideal in $k[X]$ that is both prime and principal, we readily see that M is generated by an irreducible polynomial, and hence, by Lemma (3.1), that M is maximal. It is useful to know that we do not need to find a generator for M in order to show that it is maximal.

(4.1) Theorem *If k is a discrete field, then nonzero proper prime ideals in $k[X]$ are maximal.*

Proof. Let P be a proper prime ideal in $k[X]$, and g a nonzero element of P. Let I be an ideal containing P, and f an element of I that is not in P. We proceed by induction on the degree of g. Let h be the greatest common divisor of f and g; then $h = rf + sg$ for some r,s in $k[X]$. If the degree of h is 0, then $1 = h^{-1}(rf + sg) \in I$, and therefore $I = k[X]$; so we may assume that the degree of h is greater than 0. Now h cannot be in P, as h divides f and f does not belong to P. Since P is prime, we must have $g/h \in P$, and the proof is completed by induction on the degree of g. □

(4.2) Corollary *Let k be a discrete field, and let E_1 and E_2 be discrete fields containing k. If $s_1 \in E_1$ and $s_2 \in E_2$ are algebraic over k, then the following conditions are equivalent.*

(i) *For all $p \in k[X]$, if $p(s_1) = 0$, then $p(s_2) = 0$.*

(ii) *For all $p \in k[X]$, if $p(s_2) = 0$, then $p(s_1) = 0$.*

Proof. The sets $P_i \equiv \{p \in k[X] : p(s_i) = 0\}$ are nonzero proper prime ideals in $k[X]$. Therefore they are maximal ideals, by (4.1); so if $P_i \subset P_j$, then $P_i = P_j$. □

(4.3) Lemma *Let k be a countable discrete field, and let E_1 and E_2 be discrete fields containing k. Let f be a polynomial with coefficients in k, such that $f(X) = (X-r_1)\cdots(X-r_n)$ with $r_i \in E_1$ for each i. Suppose there exists s in E_2 such that $f(s) = 0$. If* LLPO *holds, then there exists i such that for each polynomial p in $k[X]$, $p(s) = 0$ if and only if $p(r_i) = 0$.*

Proof. Let $p_1, p_2, \ldots$ be an enumeration of the polynomials in $k[X]$. Define n binary sequences $\alpha^1,\ldots,\alpha^n$ by setting $\alpha^i(j) = 1$ if and only if $p_j(r_i) = 0$ and $p_j(s) \neq 0$. Let A_i be the assertion that $\alpha^i(j) = 1$ for some j, and suppose that $A_1 \wedge A_2 \wedge \cdots \wedge A_n$ holds. Then we can construct positive integers $j(1),\ldots,j(n)$ such that $p_{j(i)}(r_i) = 0$ and $p_{j(i)}(s) \neq 0$ for each i; so f divides $g \equiv p_{j(1)}p_{j(2)}\cdots p_{j(n)}$, and $g(s) \neq 0$. But this contradicts the assumption that $f(s) = 0$. Hence, by LLPO, we have $\neg A_i$ for some i. But this says that for all $p \in k[X]$, if $p(r_i) = 0$, then $p(s) = 0$. Corollary (4.2) then yields the desired result. □

Let k be a countable discrete field, and $f \in k[X]$. Recall that a **splitting field** for f over k is a discrete extension field E of k that is generated over k by the roots of f, such that f factors into linear factors in $E[X]$. Such a field is also countable. If E_1 and E_2 are two discrete fields containing k, then an **isomorphism of E_1 with E_2 over k**, is an isomorphism of E_1 with E_2 whose restriction to k is the identity map.

(**4.4**) **Theorem** *Let k be a countable discrete field, $f \in k[X]$, and E_1 and E_2 splitting fields for f over k. If* LLPO *holds, then there is an isomorphism of E_1 with E_2 over k.*

Proof. We may assume that the degree of f is at least 2. By (4.3), we can find roots $s_1 \in E_1$ and $s_2 \in E_2$ of f such that for all $p \in k[X]$, $p(s_1) = 0$ if and only if $p(s_2) = 0$. Then the map from $k(s_1)$ to $k(s_2)$ that takes $p(s_1)$ to $p(s_2)$ is a well-defined isomorphism. If we identify $k(s_1)$ with $k(s_2)$ under this isomorphism, we can replace k by $k(s_1)$, and f by $f/(X-s_1)$. Induction on the degree of f then completes the proof. □

It follows from Theorem (4.4) that if LLPO holds, then any two algebraic closures of a countable discrete field k are isomorphic over k. The converse of Theorem (4.4) states that uniqueness of splitting fields implies LLPO. We shall prove a more informative local version.

If S is a detachable subset of $\mathbb{N}$ that contains at most one element, define the restrictions of LPO and LLPO to S as follows.

LPO_S : there exists n such that $S \subset \{n\}$.

LLPO_S: for each detachable subset A of $\mathbb{N}$, either $S \subset A$ or $S \subset \mathbb{N}\backslash A$.

It is easily seen that LPO is equivalent to LPO_S holding for each S, and that LLPO is equivalent to LLPO_S holding for each S.

(**4.5**) **Theorem** *If S is a detachable subset of $\mathbb{N}$ containing at most one element, then there exists a countable subfield k of the Gaussian number field $\mathbb{Q}(i)$ such that*

(i) *k is factorial if and only if LPO_S holds;*

(ii) *every splitting field of $X^2 + 1$ over k is isomorphic to $\mathbb{Q}(i)$ over k if and only if LLPO_S holds.*

Proof. Let $k \equiv \bigcup_{n=1}^{\infty} Q_n$, where

$$Q_n = \mathbb{Q} \quad \text{if } n \notin S,$$
$$= \mathbb{Q}(i) \quad \text{if } n \in S.$$

Then k is a countable subfield of $\mathbb{Q}(i)$, and $k = \mathbb{Q}(i)$ if and only if S is nonvoid. If LPO_S holds, then either $k = \mathbb{Q}$ or $k = \mathbb{Q}(i)$, so k is factorial. Conversely, if k is factorial, then either $X^2 + 1$ is irreducible over k, in which case S is empty; or else $X^2 + 1$ factors over k, in which case $S = \{n\}$ for some n.

Suppose that $LLPO_S$ holds, and that K is a splitting field for $X^2 + 1$ over k. Let $\alpha \in K$ satisfy $\alpha^2 = -1$, and let $A \equiv \{n : n \in S \text{ and } \alpha = i\}$. Map $K = k[\alpha]$ isomorphically onto $\mathbb{Q}(i)$ over k by taking α to i if $S \subset A$, and α to $-i$ if $S \subset \mathbb{N}\backslash A$. Conversely, suppose that every splitting field for $X^2 + 1$ over k is isomorphic to $\mathbb{Q}(i)$ over k. Let A be a detachable subset of $\mathbb{N}$, and construct a splitting field E for $X^2 + 1$ over k as follows. Define an equality on the ring $k[X]/(X^2 + 1)$ by setting $i = X$ if $S \cap A$ is nonempty, and $i = -X$ if $S\backslash A$ is nonempty; the result is the desired countable discrete field E. (Note that if an element p of $k[X]$ has a coefficient in $\mathbb{Q}(i)\backslash\mathbb{Q}$, then $i \in k$, so S is nonvoid, and so either $S \cap A$ is nonempty or $S\backslash A$ is nonempty.) Suppose φ is an isomorphism from $\mathbb{Q}(i)$ to E over k. Then $\varphi(i)$ is either equal to X or equal to $-X$. In the former case, $S \subset A$; in the latter case, $S \subset \mathbb{N}\backslash A$. □

It can be shown that if k is a factorial field, and α is **separable** - that is, α satisfies a polynomial over k that is relatively prime to its formal derivative - then $k(\alpha)$ is also a factorial field. It follows from Theorem (4.1) that if k is a factorial field of characteristic 0 (so that every element that is algebraic over k is separable), then any two algebraic closures of k are isomorphic over k. In recursive algebra, the converse is true: if any two algebraic closures of k are isomorphic over k, then k is factorial. If we could prove this converse in BISH, then by Theorem (4.5), we would be able to prove that $LLPO_S$ implies LPO_S; this is unlikely in view of topos models, constructed by Blass and Scedrov, in which $LLPO_S$ does not imply LPO_S.

We now return briefly to the context of RUSS. By modifying part of the argument in the proof of Theorem (4.5), we shall obtain a counterexample in RUSS to the statement that, given sequences of fields k_n, polynomials p_n, and splitting fields E_n and F_n of p_n over k_n, we can construct a sequence of isomorphisms $\varphi_n : E_n \to F_n$ over k_n. To do so, we need one more definition.

Two countable disjoint subsets A and B of $\mathbb{N}$ are said to be **separable** if there exists a detachable subset S of $\mathbb{N}$ that contains A and is disjoint from B. The recursive counterpart to a Brouwerian example using LLPO is a pair of **recursively inseparable sets.** Assuming CPF, and using the conventions of Chapter 3, we shall show that the sets

$$A = \{n \in \mathbb{N} : \varphi_n(n) = 0\},$$
$$B = \{n \in \mathbb{N} : \varphi_n(n) = 1\}$$

are inseparable; this is a standard result of recursive function theory. In fact, we prove the more positive statement:

(4.6) Theorem *Within* RUSS, *if* S *is a detachable subset of* $\mathbb{N}$, *then either* $S \cap B$ *is nonempty, or* $A \setminus S$ *is nonempty.*

Proof. Let S be a detachable subset of $\mathbb{N}$, and choose n so that φ_n is the characteristic function of S. If $\varphi_n(n) = 1$, then $n \in S \cap B$; if $\varphi_n(n) = 0$, then $n \in A \setminus S$. □

Using Theorem (4.6), we can construct a sequence of subfields k_n of $\mathbb{Q}(i)$, and a sequence of splitting fields K_n of $X^2 + 1$ over k_n, such that we cannot construct a sequence of isomorphisms of K_n with $\mathbb{Q}(i)$ over k_n. As in Theorem (4.6), we phrase this more positively.

(4.7) Corollary *Within* RUSS, *there exist sequences of countable discrete fields* $k_n \subset \mathbb{Q}(i)$, *and splitting fields* K_n *of* $X^2 + 1$ *over* k_n, *such that if, for each* n, λ_n *is an isomorphism of* K_n *with* $\mathbb{Q}(i)$, *then there exist* n *and* $x \in k_n$ *such that* $\lambda_n(x) \neq x$.

Proof. The reader may verify that the sets A and B of (4.6), and hence $A \cup B$, are countable. Let $x_1, x_2, \ldots$ be an enumeration of $A \cup B$, and for all positive integers n,j define

$$k_{n,j} \equiv \mathbb{Q} \quad \text{if } n \neq x_j,$$
$$\equiv \mathbb{Q}(i) \quad \text{if } n = x_j.$$

Then $k_n \equiv \bigcup_{j=1}^{\infty} k_{n,j}$ is a countable discrete field such that $k_n = \mathbb{Q}(i)$ if $n \in A \cup B$, and $k_n = \mathbb{Q}$ otherwise. Define an equality on $k_n[X]/(X^2 + 1)$ by setting $i = X$ if $n \in A$, and $i = -X$ if $n \in B$. The resulting field K_n is a splitting field for $X^2 + 1$ over $\mathbb{Q}(i)$. Suppose that for each n, λ_n is an isomorphism from K_n to $\mathbb{Q}(i)$. Then as

$$\lambda_n(X)^2 + 1 = \lambda_n(X^2 + 1) = 0,$$

we see that either $\lambda_n(X) = i$ or $\lambda_n(X) = -i$. The set

$$S = \{n \in \mathbb{N} : \lambda_n(X) = i\}$$

is detachable from $\mathbb{N}$. So either there exists n in $S \cap B$, in which case $\lambda_n(-i) = \lambda_n(X) = i$; or else there exists n in $A \backslash S$, in which case $\lambda_n(i) = \lambda_n(X) = -i$. □

5. Finitely presented modules

In the rest of this chapter, we shall use the term 'map' in its categorical sense – that is, a **map** between two algebraic structures will be a structure-preserving mapping.

A general method for presenting an algebraic structure S is to specify a set of generators and a set of relations that hold among those generators. The elements of S are constructed by formal repeated application of the algebraic operations to the generators; two elements of S are equal if their equality is a consequence of the given relations. For example, we can present an abelian group by specifying the set of generators $x_0, x_1, \ldots$ and the relations

$$px_0 = 0 \text{ and } p^n x_n = x_0 \text{ for } n > 0,$$

where p is a fixed prime.

A **finite presentation** consists of a finitely enumerable set of generators and a finitely enumerable set of relations. A finite presentation of an abelian group is specified by a matrix of integers: any relation holding among generators $x_1, \ldots, x_n$ of an abelian group can be put in the form

$$a_1x_1 + a_2x_2 + \cdots + a_nx_n = 0,$$

where $(a_1, \ldots a_n) \in \mathbb{Z}^n$. Similarly, a finite presentation of a (left) module M over a ring R is given by an m-by-n matrix A over R. The row $(a_{i1}, \ldots, a_{in})$ of A corresponds to the relation $\Sigma_{j=1}^{n} a_{ij}x_j = 0$, where $x_1, \ldots, x_n$ are the given generators of M; the rows of A are elements of R^n, which is a left R-module under coordinatewise operations. The relations that are **consequences** of the given relations correspond to elements of the submodule of R^n generated by the rows of A, and constitute all the relations satisfied by the given generators. In other words, an R-module

M is **finitely presented** if there is a map φ from a finite-rank free R-module R^n onto M such that the kernel of φ is finitely generated; in the above notation, $\varphi(r_1,\ldots,r_n) = \Sigma_{j=1}^n r_jx$. We shall use the term 'finite presentation' for such a map.

Classically, *every* subgroup of $\mathbb{Z}^n$ is finitely generated, so abelian groups are finitely presented if and only if they are finitely generated.

The fundamental theorem of abelian group theory is that a finitely presented abelian group is a direct sum of finite and infinite cyclic groups; the proof consists of reducing the presentation matrix, by elementary row and column operations, to Smith normal form. The finitely presented modules over a discrete field are precisely the finite-dimensional vector spaces: to show this, given a finitely presented module, we extract a basis by reducing the presentation matrix to row echelon form and choosing the generators corresponding to the columns of the leading nonzero entries of the rows.

It is a remarkable fact that if a module M is finitely presented, then the set of relations of any finite generating set is finitely generated; that is, if one map of R^n onto M has a finitely generated kernel, then any map of R^m onto M has a finitely generated kernel.

(5.1) Theorem *If M is a finitely presented R-module, and f is a map from R^m onto M, then the kernel of f is finitely generated.*

Proof. Let g be a map of R^n onto M such that the kernel K_g of g is finitely generated, and let K_f denote the kernel of f. We shall use **Schanuel's trick** to show that $K_f \oplus R^n$ is isomorphic to the finitely generated module $K_g \oplus R^m$; from which it follows that K_f is finitely generated.

Construct maps $\varphi_f:R^m \to R^n$ and $\varphi_g:R^n \to R^m$, such that $g\varphi_f = f$ and $f\varphi_g = g$, by defining φ_f amd φ_g on the finite bases of R^m and R^n so that the equations hold on the bases; this is possible because both g and f are onto. Map $K_f \oplus R^n$ to $K_g \oplus R^m$ by taking (k_f,r_n) to (k_g,r_m) where

$$k_g \equiv \varphi_f(k_f) + r_n - \varphi_f\varphi_g(r_n),$$
$$r_m \equiv \varphi_g(r_n) - k_f.$$

and map $K_g \oplus R^m$ to $K_f \oplus R^n$ by setting

$$k_f \equiv \varphi_g(k_g) - r_m + \varphi_g\varphi_f(r_m),$$
$$r_n \equiv \varphi_f(r_m) + k_g.$$

It is easily checked that these maps are inverses of each other. □

A ring is **coherent** if every finitely generated left ideal is finitely presented; a module is **coherent** if every finitely generated submodule is finitely presented. Using Theorem (5.1), we see that a module M over a ring R is coherent if and only if every map from a finite-rank free R-module into M has a finitely generated kernel. Clearly, every discrete field is coherent; as is the ring of integers $\mathbb{Z}$, since every finitely generated ideal admits a presentation with a single generator x, and either no relations or the single relation $x = 0$.

Coherence is a critical property from the computational point of view: it allows us to compute generators for the intersection of finitely generated submodules, and it allows us to compute generators for the annihilators of elements.

(5.2) Theorem *If M is a coherent R-module, then the intersection of any two finitely generated submodules of M is finitely generated; and if $x \in M$, then $\{r \in R : rx = 0\}$ is a finitely generated left ideal of R.*

Proof. Let A_1 and A_2 be finitely generated submodules of M; then as M is coherent, A_1 and A_2 are finitely presented. For $i = 1,2$, let f_i be a map of a finite-rank free R-module F_i onto A_i. Let K be the kernel of the induced map $f: F_1 \oplus F_2 \to M$ defined by

$$f(x_1,x_2) \equiv f_1(x_1) + f_2(x_2).$$

Then $f(F_1 \oplus F_2) = A_1 + A_2$ is finitely generated and therefore finitely presented (as M is coherent); whence, by (5.1), K is finitely generated. Let e be the projection of $F_1 \oplus F_2$ onto F_1; we shall show that $A_1 \cap A_2 = f_1eK$, from which it follows that $A_1 \cap A_2$ is finitely generated, as K is finitely generated. If $x \in A_1 \cap A_2$, then there exist y_1,y_2 such that $x = f_1(y_1) = f_2(y_2)$, and therefore $x = f_1e(y_1,-y_2)$. Conversely, if $(y_1,-y_2)$ is in K, then $f(y_1) = f(y_2)$ is in $A_1 \cap A_2$; whence $f_1e(y_1,-y_2) = f(y_1)$ is in $A_1 \cap A_2$.

For the second claim, the cyclic submodule Rx of M is finitely presented because M is coherent; the left ideal $\{r \in R : rx = 0\}$ is the kernel of the map from R to M that takes r to rx, and so is finitely generated, by (5.1). □

Theorem (5.2) actually characterizes coherence, but we will not need this fact.

The computational content of coherence is already significant in the case of rings of polynomials in several variables over a discrete field: it is nontrivial to pass from generators for ideals I and J in such a ring to generators for $I \cap J$. For polynomials in *one* variable we effect this passage by using the Euclidean algorithm to compute single generators for the two ideals, and then computing the least common multiple of these two generators. The natural way to approach the problem of showing that polynomial rings in several variables are coherent is to show that coherence is preserved upon passing from R to $R[X]$; however this is false, even classically. What we *can* do is consider rings R that are both coherent and Noetherian (in a sense to be defined in Section 6), and show that these properties are inherited by $R[X]$.

We will need the fact that the finite-rank free modules R^n over a coherent ring R are coherent. Translated into more mundane terms, this says that if we are given an m-by-n matrix (a_{ij}) with entries in R, then we can find a general solution to the system of equations $\Sigma_{i=1}^{m} x_i a_{ij} = 0$, in the sense that we can find a finite number of solutions such that any solution is a linear combination of those solutions. (The variables are on the left because we are dealing with left modules, so matrices representing homomorphisms act on the right). This follows immediately from the following more general fact.

(5.3) Theorem *If* A *and* B *are coherent* R*-modules, then so is* $A \oplus B$.

Proof. Let S be a finitely generated submodule of $A \oplus B$, let f map R^m onto S, and let β be the projection of $A \oplus B$ onto B. As B is coherent, the finitely generated submodule βS of B is finitely presented; whence, by (5.1), βf has a finitely generated kernel K. Thus $A \cap S = fK$ is finitely generated and therefore finitely presented. Let $f_1 : F_1 \to A \cap S$ and $f_2 : F_2 \to \beta S$ be finite presentations. Construct a map $\varphi : F_1 \oplus F_2 \to S$ that agrees with f_1 on F_1, such that $f_2 = \beta\varphi$ on F_2. As $\beta f_1 = 0$ we have

$$\beta\varphi(x,y) = \beta f_1(x) + f_2(y) = f_2(y).$$

Thus if $\varphi(x,y) = 0$, then $f_2(y) = 0$; while if $f_2(y) = 0$, then $\varphi(0,y) = 0$. Hence if e is the projection of $F_1 \oplus F_2$ onto F_2, then $e(\mathrm{Ker}\ \varphi) = \mathrm{Ker}\ f_2$.

Also, trivially, $F_1 \cap \operatorname{Ker} \varphi = \operatorname{Ker} f_1$. Since $\operatorname{Ker} f_1$ and $\operatorname{Ker} f_2$ are finitely generated, by (5.1), we see that $\operatorname{Ker} \varphi$ is finitely generated. Therefore S is finitely presented. □

6. Noetherian rings

Das ist nicht Mathematik. Das ist Theologie. (P. Gordan)

The seminal nonconstructive argument in algebra is Hilbert's proof that every ideal in the ring of polynomials in several variables over a field is finitely generated (*Über die Theorie der algebraischen Formen*, Math. Ann. 36(1890), 473–534). In applying this theorem to Gordan's problem of finding finite sets of generators for certain rings of invariant forms, Hilbert reduced the problem to that of finding finite sets of generators for certain ideals. As the rings and associated ideals in Gordan's problem are described in a finitistic way, Gordan expected an explicit description of the generators, which he had provided for the two variable case. Hilbert's proof that *all* ideals are finitely generated had to be nonconstructive, given the latitude one has in specifying ideals; what disturbed the mathematicians of his day was its application to a problem that invited a computational solution.

Nowadays the Hilbert basis theorem is taken to be

If R is a Noetherian ring, then so is the ring of polynomials $R[X]$.

In fact, the standard classical proof is constructive; the problem is that the hypotheses are never satisfied. Classically, a (left) Noetherian ring may be defined to be a ring in which every left ideal is finitely generated; but even the two-element field $\mathbb{Z}_2$ fails to satisfy this definition from a constructive point of view! Indeed, if we take the ideal I of $\mathbb{Z}_2$ generated by a binary sequence, then a finite set of generators for I will tell us whether or not the sequence has a term equal to 1.

To obtain a useful constructive definition of a Noetherian ring, we must find a suitable version of the ascending chain condition on left ideals. The **classical ascending chain condition** on a partially ordered set S says that if $s_1 \leq s_2 \leq s_3 \leq \cdots$ is an increasing sequence in S, then there exists n such that $s_m = s_n$ for each $m \geq n$. The field $\mathbb{Z}_2$ fails to satisfy the classical ascending chain condition on ideals (partially orded by inclusion), as can be seen by taking I_j to be the ideal generated by the first j terms of some binary sequence. The correct constructive definition of the **ascending chain condition** requires only that there exist n such that $s_n = s_{n+1}$: it suffices to find a place where the chain *pauses*, rather than a place where the chain *stabilizes*. Classically, this is equivalent to the classical ascending chain condition.

We say that a ring is (left) **Noetherian** if it satisfies the ascending chain condition on finitely generated left ideals (partially ordered by inclusion). More generally, we say that an R-module is **Noetherian** if it satisfies the ascending chain condition on finitely generated submodules. We must require that the left ideals be finitely generated, which is classically equivalent to allowing them to be arbitrary, in order to get a hold on them: to appreciate this, let (a_{ij}) be a doubly indexed binary sequence such that $(a_{ij})_{j=1}^{\infty}$ is increasing for each i, and consider the sequence of ideals I_j in $\mathbb{Z}_2$ generated by the sets $\{a_{ij} : i \in \mathbb{N}^+\}$.

We can show that the ring $\mathbb{Z}$ of integers is Noetherian as follows. Suppose $I_1 \subset I_2 \subset \ldots$ is an ascending chain of finitely generated ideals of $\mathbb{Z}$, and use the Euclidean algorithm to construct, for each j, a nonnegative generator m_j for I_j. As $I_j \subset I_{j+1}$, it follows that m_{j+1} divides m_j; thus for some $j \leq m_2 + 1$ we must have $m_j = m_{j+1}$, so that $I_j = I_{j+1}$.

A similar argument shows that the polynomial ring $k[X]$ over a discrete field k is Noetherian. For polynomial rings in several variables over a discrete field, or over $\mathbb{Z}$, the argument is much more complicated. It would suffice, of course, to prove the Hilbert basis theorem; but no one has yet succeeded in giving a constructive proof of the Hilbert basis theorem, as previously stated, using our definition of Noetherian.

We can put together an argument for the Hilbert basis theorem for countable discrete R by invoking Markov's principle and using some classical reasoning. The following argument can be transformed into a *bona fide* proof within recursive algebra, but is unacceptable to any of the constructive schools. Given an ascending chain $I_1 \subset I_2 \subset \ldots$ of finitely generated ideals of R, we systematically attempt to write each of the generators of I_{j+1} as a linear combination of the generators of I_j; we can do this, by a dovetailing argument, simultaneously for all j. Because there is a classical argument that proves $R[X]$ Noetherian, there must exist j such that $I_j = I_{j+1}$; Markov's principle then guarantees that we can find a way to write the generators of I_{j+1} in terms of the generators of I_j.

If we want a truly constructive proof, we must postulate some additional property enjoyed by rings of polynomials over discrete fields. The outstanding candidate is coherence: we saw in Theorem (5.2) how coherence allows us to construct generators for intersections of finitely generated submodules; moreover, the assumption of coherence is classically redundant - every Noetherian ring R is coherent, as every submodule of R^m is finitely generated - so theorems proved with its aid have the same classical scope as those proved without.

Before proving the fundamental lemma about finitely generated ideals in polynomial rings over coherent Noetherian rings, we verify that if R is Noetherian, then so is R^m. This is a consequence of the following standard theorem, whose proof is more complicated than its classical counterpart because our sequences pause rather than stabilize, and our submodules must be finitely generated.

(6.1) Theorem *If A and B are Noetherian modules, then so is* $A \oplus B$.

Proof. Let β be the projection of $A \oplus B$ on B, and let $S_1 \subset S_2 \subset \cdots$ be a chain of finitely generated submodules of $A \oplus B$. Then $\beta S_1 \subset \beta S_2 \subset \cdots$ is a chain of finitely generated submodules of B, so there exists n such that $\beta S_n = \beta S_{n+1}$. Therefore $S_{n+1} = S_n + F$ for some finitely generated submodule F of A (the submodule F is generated by elements $x_i - y_i$, where the x_i generate S_{n+1}, the y_i are in S_n, and $\beta x_i = \beta y_i$). Repeating this construction, we obtain a strictly increasing sequence of positive integers $n(i)$, and an ascending chain of finitely generated submodules F_i of A, such that $S_{n(i)+1} = S_{n(i)} + F_i$ for each i (note the countable choice, as the F_i are not uniquely specified). As A is Noetherian, there exists i such that $F_i = F_{i-1} \subset S_{n(i)}$; whence $S_{n(i)+1} = S_{n(i)}$. □

If R is a ring, we denote by $R[X]_n$ the R-module $\{f \in R[X] : \deg f < n\}$. As $R[X]_n$ is isomorphic to R^n under the map taking $a_0 + a_1X + \cdots + a_{n-1}X^{n-1}$ to $(a_0,\ldots,a_{n-1})$, we see from Theorems (5.3) and (6.1) that if R is coherent (respectively, Noetherian), then so is $R[X]_n$. Note that if M is an R-submodule of $R[X]_n$, then X^mM is an R-submodule of $R[X]_{n+m}$.

The next lemma reduces questions about finitely generated left ideals in $R[X]$ to questions about finitely generated R-submodules of $R[X]$.

(6.2) Lemma *Let R be a coherent Noetherian ring, and I the left ideal of $R[X]$ generated by $f_1,\ldots,f_s$. If $f_i \in R[X]_n$ for each i, then there exists a finitely generated R-module $M \subset R[X]_n$ such that $XM \cap R[X]_n \subset M$, and $I \cap R[X]_m = \Sigma_{i=0}^{m-n} X^i M$ for each $m \geq n$.*

Proof. Construct a chain $N_1 \subset N_2 \subset \cdots$ of finitely generated submodules of $I \cap R[X]_n$ as follows. Let $N_1 \equiv Rf_1 + \cdots + Rf_s$, and for each positive integer k let $N_{k+1} \equiv N_k + XN_k \cap R[X]_n$. As $R[X]_{n+1}$ is coherent, the modules N_k are finitely generated; as $R[X]_n$ is Noetherian there exists k such that $N_k = N_{k+1}$. Setting $M \equiv N_k$, we clearly have $XM \cap R[X]_n \subset M$.

Consider any $m \geq n$. As $M \subset I \cap R[X]_n$, we have $\Sigma_{i=0}^{m-n} X^i M \subset I \cap R[X]_m$. To show that $I \cap R[X]_m \subset \Sigma_{i=0}^{m-n} X^i M$, suppose that $f \in I \cap R[X]_m$. Writing $f = \Sigma_{i=1}^{s} g_i f_i$, where $g_i \in R[X]_d$ for each i, we proceed by induction on d (for all $m \geq n$). If $d = 1$, then $f \in N_1 \subset M$ and we have finished. If $d > 1$, define $h_i \in R[X]$ by $g_i = g_i(0) + Xh_i$, set $f^* \equiv \Sigma_{i=1}^{s} h_i f_i \in I$, and note that $h_i \in R[X]_{d-1}$. Then $f = \Sigma_{i=1}^{s} g_i(0) f_i + Xf^*$; so that $Xf^* \in R[X]_m$, and therefore $f^* \in R[X]_{m-1} \subset R[X]_m$. If $m = n$, then our induction hypothesis shows that $f^* \in M$; so that $Xf^* \in XM \cap R[X]_n \subset M$, whereupon $f \in N_1 + M = M$. If $m > n$, then our induction hypothesis shows that $f^* \in \Sigma_{i=0}^{m-1-n} X^i M$; whence

$$f \in N_1 + X\Sigma_{i=0}^{m-1-n} X^i M \subset \Sigma_{i=0}^{m-n} X^i M. \quad \square$$

(6.3) Corollary *Let R be a coherent Noetherian ring, and k a positive integer. If I is a finitely generated left ideal of $R[X]$, then $I \cap R[X]_k$ is a finitely generated R-module. In particular, $I \cap R$ is a finitely generated left ideal of R.*

Proof. Let $n \geq k$, and construct $M \equiv I \cap R[X]_n$ as in (6.2). Then $I \cap R[X]_k = M \cap R[X]_k$ is finitely generated, by (5.2), because $R[X]_n$ is coherent and both M and $R[X]_k$ are finitely generated. $\square$

A map φ from a ring R to a ring S **reflects finitely generated left ideals** if whenever I is a finitely generated left ideal of S, then $\varphi^{-1}I$ is a finitely generated left ideal of I. The second part of Corollary (6.3) shows that if R is a coherent Noetherian ring, then the natural map $\varphi: R \to R[X]$ reflects finitely generated left ideals; for in that case, $\varphi^{-1}I = I \cap R$.

We now prove the first part of the Hilbert basis theorem.

(6.4) Lemma *If R is a coherent Noetherian ring, then $R[X]$ is a coherent ring.*

Proof. Let I be a finitely generated left ideal of $R[X]$, and let $f \in R[X]$. Choose n such that $R[X]_n$ contains f and a finite set of generators of I. By (6.2), $M \equiv I \cap R[X]_n$ is finitely generated. Choose a finite set G of generators of M containing a finite set G' of generators of $M \cap R[X]_{n-1}$, and also containing XG'. By the choice of n, the R-submodule M generates I as left $R[X]$-ideal, so G generates I as a left $R[X]$-ideal. Construct a finite set U of $R[X]$-linear relations among the elements of G as follows: as $R[X]_n$ is a coherent R-module, the set of R-linear relations among the elements of G has a finite set of generators; to obtain U, adjoin to these generators the $R[X]$-linear relations $X \cdot g = Xg$ for each $g \in$ G'.

Given an $R[X]$-linear relation $\Sigma r_g g = 0$, we shall show that it is a consequence of U. Choosing m so that $\deg r_g \leq m$ for each $g \in$ G, we proceed by induction on m. If $m = 0$, then our relation is R-linear and therefore a consequence of U. If $m > 0$, write $r_g = s_g + a_g X^m$, where $a_g \in R$ and $\deg s_g < m$. From the equation $\Sigma s_g g + X^m \Sigma a_g g = 0$ and the fact that G $\subset R[X]_n$, we see that

$$\deg \Sigma a_g g = \deg \Sigma s_g g - m < n.$$

Hence $\Sigma a_g g \in M \cap R[X]_{n-1}$; so that $\Sigma_{g \in G}\, a_g g = \Sigma_{g \in G'}\, b_g g$, where each b_g belongs to R. This is an R-linear relation, so we can use U to replace $\Sigma_{g \in G}\, a_g g$ by $\Sigma_{g \in G'}\, b_g g$, thereby yielding the relation

$$\Sigma_{g \in G}\, s_g g + X^m \Sigma_{g \in G'}\, b_g g = 0.$$

Now replace $X \cdot g$ by Xg for $g \in$ G'. The resulting relation is

$$\Sigma_{g \in G}\, s_g g + X^{m-1} \Sigma_{g \in G'}\, b_g (Xg) = 0.$$

By induction on m, this relation is a consequence of U. Thus I is finitely presented. □

If R is a ring, and I is a left ideal of $R[X]$, define

$$L(I) \equiv \{a_n \in R : a_n X^n + a_{n-1} X^{n-1} + \cdots + a_0 \in I\}$$

to be the set of leading coefficients of the polynomials in I. Note that a polynomial may have different leading coefficients (one of them zero), depending on how it is written, and that $L(I)$ is a left ideal of R. The ideal $L(I)$ plays a rôle in all treatments of the Hilbert basis theorem. It is crucial that $L(I)$ be finitely generated; this poses no problem in the classical treatment because *all* ideals of R are assumed to be finitely generated, but some work is required to establish the constructive result.

(6.5) Lemma *Let R be a coherent Noetherian ring, and I a finitely generated left ideal of $R[X]$. Then the left ideal $L(I)$ of R is finitely generated. Let $J \supset I$ be a left ideal of $R[X]$ such that $L(I) = L(J)$, and let m be a positive integer. If $I \cap R[X]_m$ generates I, then $J \cap R[X]_m$ generates J.*

Proof. Let n and $M = I \cap R[X]_n$ be as in (6.2), and let $L_n(I) \subset L(I)$ be the set of leading coefficients of polynomials in M. As $XM \cap R[X]_n \subset M$, the set $L_n(I)$ is the image of the map taking each polynomial in M to its coefficient of X^{n-1}. As M is finitely generated, so is $L_n(I)$. We complete the proof of the first claim of our lemma by showing that $L_n(I) = L(I)$. Suppose $f \equiv f_{m-1}X^{m-1} + \cdots + f_1X + f_0 \in I \cap R[X]_m$. If $m \leq n$, then $f_{m-1} \in L_n(I)$. If $m > n$, then $f \in \Sigma_{i=0}^{m-n} X^iM$, by our choice of M, and so we can find $g \in M$ with leading coefficient f_{m-1}; whence $f_{m-1} \in L_n(I)$.

Now suppose that $J \supset I$ is a left ideal of R such that $L(I) = L(J)$, and let m be a positive integer such that $I \cap R[X]_m$ generates I. By rechoosing n and M, if necessary, we may assume that $n = m$. Let $f \equiv f_{d-1}X^{d-1} + \cdots + f_1X + f_0 \in J \cap R[X]_d$; we shall show by induction on d that f is in the left ideal generated by $J \cap R[X]_n$. If $d \leq n$, then $f \in J \cap R[X]_n$. If $d > n$, then as $L_n(I) = L(I) = L(J)$, we can find $g \equiv g_{n-1}X^{n-1} + \cdots + g_1X + g_0 \in M$ with $g_{n-1} = f_{d-1}$. Then $f - X^{d-n}g$ belongs to $J \cap R[X]_{d-1}$ and therefore, by our induction hypothesis, to the left ideal generated by $J \cap R[X]_n$. As $g \in M \subset J \cap R[X]_n$, it follows that f is in the left ideal generated by $J \cap R[X]_n$. □

Finally we are in a position to prove a constructively useful version of the **Hilbert basis theorem:**

(6.6) Theorem *If R is a coherent Noetherian ring, then so is $R[X]$.*

Proof. By Lemma (6.4), $R[X]$ is coherent; so it remains to show that $R[X]$ is Noetherian. Let $I_1 \subset I_2 \subset \cdots$ be a chain of finitely generated left ideals of $R[X]$. With $\nu(1) \equiv 1$, construct a strictly increasing sequence

$\upsilon(1), \upsilon(2), \ldots$ of positive integers, and a sequence $n(1), n(2), \ldots$ of nonnegative integers, such that for each $m \geq 1$,

(i) $I_{\upsilon(m)} \cap R[X]_{n(m)}$ generates $I_{\upsilon(m)}$,

(ii) $I_{\upsilon(m)} \cap R[X]_{n(m-1)} = I_{\upsilon(m)+1} \cap R[X]_{n(m-1)}$.

Indeed, having constructed $\upsilon(m-1)$ and $n(m-1)$, as $R[X]_{n(m-1)}$ is Noetherian and $I_i \cap R[X]_{n(m-1)}$ is finitely generated (by (6.3)), we can choose $\upsilon(m) > \upsilon(m-1)$ so that (ii) obtains; we then choose $n(m)$ so that (i) is satisfied. The left ideals $L(I_{\upsilon(1)}), L(I_{\upsilon(3)}), L(I_{\upsilon(5)}), \cdots$ are finitely generated, by (6.5); as $L(I_{\upsilon(1)}) \subset L(I_{\upsilon(3)}) \subset \ldots$, we can find m such that $L(I_{\upsilon(m-1)}) = L(I_{\upsilon(m+1)})$. Then also $L(I_{\upsilon(m)}) \subset L(I_{\upsilon(m+1)}) = L(I_{\upsilon(m-1)})$, and so $L(I_{\upsilon(m-1)}) = L(I_{\upsilon(m)})$. But $I_{\upsilon(m-1)} \cap R[X]_{n(m-1)}$ generates $I_{\upsilon(m-1)}$, so Lemma (6.5) tells us that $I_{\upsilon(m)} \cap R[X]_{n(m-1)}$ generates $I_{\upsilon(m)}$. Similarly $I_{\upsilon(m)+1} \cap R[X]_{n(m-1)}$ generates $I_{\upsilon(m)+1}$. Thus, by (ii), $I_{\upsilon(m)} = I_{\upsilon(m)+1}$. □

It follows from Theorem (6.6) that if k is either a discrete field or the ring of integers, then the polynomial ring $R = k[X_1,\ldots,X_n]$ is a coherent Noetherian ring. This result embodies a fair amount of computational content: for example, it follows from Theorem (5.2) that, given polynomials $f_1,\ldots,f_s,g_1,\ldots,g_t$ in R, we can find polynomials $h_1,\ldots,h_u$ that generate the intersection of the ideals in R generated by $f_1,\ldots,f_s$ and by $g_1,\ldots,g_t$.

PROBLEMS.

1. **The division algorithm.** Let k be a discrete field. Show by induction on $deg\ b$ that if $a,b \in k[X]$ with $a \neq 0$, then there exist $q,r \in k[X]$ such that $b = qa + r$, and $r = 0$ or $deg\ r < deg\ a$.

2. **The Euclidean algorithm.** Let k be a discrete field. Use the division algorithm, and induction on $min(deg\ a,\ deg\ b)$ to show that if $a,b \in k[X]$, then there exist $s,t \in k[X]$ such that $sa + tb$ divides both a and b.

3. **Unique factorization.** Let k be a discrete field, and p an irreducible polynomial over k. Use the Euclidean algorithm to show that if p divides ab, then p divides a, or p divides b.

Conclude that $k[X]$ satisfies condition (ii) of a unique factorization domain.

4. Let a and b be real numbers, $c = sup(a,b)$, and $d = inf(a,b)$. Show that $(X - a)(X - b) = (X - c)(X - d)$ in $\mathbb{R}[X]$. Use this to construct a Brouwerian example showing that $\mathbb{R}[X]$ does not satisfy condition (ii) of a unique factorization domain.

5. Prove Gauss's lemma by showing that if p is a prime dividing all the coefficients of fg, then either p divides all the coefficients of f, or else p divides all the coefficients of g. (Pass to the ring $\mathbb{Z}_p$ of integers modulo p.)

6. The proof of Gauss's lemma outlined in Problem 5 works for coefficients in $\mathbb{Z}$, but fails for coefficients in $k[Y]$ because in that case, we cannot necessarily find an irreducible polynomial $p(Y)$ that divides all the coefficients of $g(X)$, even if g is not primitive. Amend the proof so that it works in that case also. (Compute greatest common divisors by the Euclidean algorithm, and use induction on the degree of the (not necessarily irreducible) polynomial p.)

7. Show that the field $\mathbb{Q}(X)$ of quotients of $\mathbb{Q}[X]$ is factorial by showing that if k is a discrete field, and if you can factor polynomials in $k[X]$, then you can factor polynomials in $k[X,Y]$. (Use Kronecker's trick of substituting $Y = X^n$ for suitably large n.)

8. Show that if $f(X)$ is an irreducible polynomial with coefficients in a discrete field k, and I is the ideal generated by $f(Y)$, then the quotient ring $E \equiv k[Y]/I$ is a discrete field containing k such that f has a root in E. Show that E is a finite-dimensional vector space over k, and construct a basis for it.

9. Construct a Brouwerian example of a discrete cyclic group that is not finitely presented.

10. Show that if M is a coherent module, and $A \subset M$ is a finitely generated submodule, then M/A is coherent.

11. Show directly, without using Lemma 6.4 or Theorem 6.6, that if k is a discrete field, then $k[X]$ is a coherent Noetherian ring.

12. Show that if A is a submodule of M, then M is Noetherian if and only if A and M/A are Noetherian.

13. An R-module M is said to **have detachable submodules** if each finitely generated submodule of M is detachable. Show that if A is a finitely generated submodule of M, then M has detachable submodules if and only if A and M/A have detachable submodules. Conclude that if R has detachable left ideals, then R^n has detachable submodules.

14. Show that if R is a coherent Noetherian ring with detachable left ideals, then so is $R[X]$. Conclude that, given polynomials $f_1,\ldots,f_n,g$ in $\mathbb{Z}[X_1,\ldots,X_m]$, we can decide whether or not g is in the ideal generated by $f_1,\ldots,f_n$.

15. Let I be a finitely generated ideal in a coherent Noetherian commutative ring R with detachable ideals. The **radical** of I consists of all $r \in R$ such that $r^n \in I$ for some $n \in \mathbb{N}$. Show that the radical of I is a detachable ideal. (Consider the ideals $J_m \equiv \{s \in R : sr^m \in I\}$.) Construct a Brouwerian example to show that the radical of I need not be finitely generated. (Let R lie between $\mathbb{Q}$ and $\mathbb{Q}[X]/(X^2)$, and let $I \equiv 0$.)

NOTES

Seidenberg refers to a factorial field as 'a field that satisfies condition F'; recursive algebraists talk about 'a field with a splitting algorithm'.

Our Brouwerian example (Theorem (2.1)) of a nonfactorial field is simpler than van der Waerden's example in *Eine Bemerkung über die Unzerlegbarkeit von Polynomen* (Math. Ann. 102 (1930), 738–739), which was the field generated by $\{\sqrt{a_n p_n} : n \in \mathbb{N}\}$, where p_n is the nth prime. The latter has the virtue of also being a counterexample in recursive algebra.

Kronecker's theorem, which appeared in *Grundzüge einer arithmetischen Theorie der algebraishen Grossen* (section 4), (J. reine angewandte Math., 92 (1882), 1–122) is attributed to von Schubert (1793) in Knuth, *Seminumerical Algorithms* (2nd edn, McGraw-Hill, 1981, p. 431), where reference is made to M. Cantor, *Geschichte der Mathematik* (Teubner, 1908, Vol. 4).

The world's simplest axiom of choice was introduced in Richman, *Seidenberg's condition P* (**Springer 873**, 1-11), in connection with the problem of constructing, for an arbitrary discrete field, a discrete extension field in which a given polynomial has a root. Fourman and Scedrov give a topos-theoretic counterexample in *The 'world's simplest axiom of choice' fails* (Manuscripta Math. 38 (1982), 325–332); see also Chapter 7.

The proof of Theorem (4.5) was suggested in the discussion following Corollary (3.9) of Julian, Mines and Richman, *Algebraic numbers, a constructive development* (Pacific J. Math. 74 (1978), 91–102).

In the unexpurgated editions of van der Waerden's *Modern Algebra* (Ungar, New York, 1953), there are sections on constructive field theory. One theorem there states that a finite-dimensional separable extension of a factorial field is factorial. This theorem may also be found in Mines and Richman, *Separability and factoring of polynomials*, (Rocky Mtn. J. Math. 12 (1982), 43–54). Seidenberg, in *Constructions in algebra* (Trans. Amer. Math. Soc. 197 (1974), 273–313), and Mines and Richman (reference above) have constructed examples showing the necessity of the separability condition. Seidenberg showed that the separability condition can be eliminated if the field k also satisfies a property he calls **condition** P: *if k has characteristic p, then we can decide whether or not any given system of linear equations over k has a nontrivial solution in $k^p = \{x^p : x \in k\}$.* If a factorial field satisfies condition P, then any finite-dimensional extension field is factorial and satisfies condition P.

The theorem of recursive algebra, that a countable discrete field is factorial if it has a unique algebraic closure, originates with Metakides and Nerode, in *Effective content of field theory* (Annals of Math. Logic 17 (1979), 289–320). Translated into our terms, their theorem says that if a discrete field k is not factorial, and E is an algebraic closure of k, then there is a algebraic closure of k that is not isomorphic to E over k. The proof can be effected, up to a point, in the context of BISH with CPF and Markov's principle. By Markov's principle, it suffices to be able to decide whether any given polynomial over k is reducible (without finding its factors). An ingenious argument using *finite injury priority methods* makes possible the construction of a function $\varphi:k[X] \to \{0,1\}$ such that the set

$$\{p : \varphi(p) = 0 \text{ and } p \text{ is irreducible, or } \varphi(p) = 1 \text{ and } p \text{ is reducible}\}$$

cannot be infinite. Resorting to classical reasoning at this point, we can modify φ at a finite number of places to get the characteristic function of the set of reducible polynomials.

In their paper *Small decidable sheaves* (J. Symbolic Logic 51 (1986), 726-731), Blass and Scedrov show that in certain topos models, LLPO_S holds but LPO_S does not. Neither Markov's principle nor Church's thesis holds in their models, so although they preclude a proof strictly within BISH, they do not rule out a proof using CPF and Markov's principle. The reader familiar with the (wildly nonconstructive) theory of ultrafilters will notice that LLPO_S is to LPO_S as an ultrafilter is to a fixed ultrafilter.

Brouwerian examples translate into sequences of examples in recursive mathematics because, classically, you can always decide one thing - for example, does $p(X)$ have a root in k? - with a (constant) recursive function, but you can't necessarily decide a sequence of things. The unprovability of LPO within BISH corresponds to the existence of recursively enumerable nonrecursive sets; the unprovability of LLPO to the existence of recursively inseparable sets.

The paper *Anneaux et modules cohérents* (J. Algebra 15 (1970), 455-472), by J-P. Soublin, contains an example of a coherent commutative ring R such that $R[X]$ is not coherent, but leaves open the question of whether we can find such an R which is a domain.

The first meaningful constructive Hilbert basis theorem was proved by Jonathan Tennenbaum in his Ph.D dissertation, supervised by Errett Bishop. His definition of Noetherian relied essentially on Bishop's notion of an *operation*, which is a finite procedure that produces an element of Y when given an element of X, but which need not be a function, as it may not return equal elements when given equal elements. This notion translates very well into RUSS, but does not fare so well in INT or in standard classical models of intuitionistic logic. Moreover, although in most applications an operation can be thought of as a function from X to the set of nonempty subsets of Y, this interpretation does not fit with Tennenbaum's definition. For a development of Tennenbaum's approach in the discrete case see Chapter VIII, Section 3, of **MRR**. It is interesting to note, in connection with Section 1, that Tennenbaum devoted the greater part of his dissertation to the nondiscrete case, and considered that the more significant contribution.

The Hilbert basis theorem in the form (6.6) is due to Richman, *Constructive aspects of Noetherian rings* (Proc. Amer. Math. Soc. 44 (1974), 436-441), and Seidenberg, *What is Noetherian?* (Rend. Sem. Mat. e Fis. di Milano, 44(1974), 55-61).

Chapter 5. Intuitionism

In which Brouwer's intuitionistic mathematics is discussed from the point of view of the mathematician, with minimal involvement in philosophical issues. The immediate mathematical consequences of Brouwer's philosophical introspection are summarised in two principles, the principle of continuous choice *and the* fan theorem, *which are then used to establish the fundamental results of intuitionistic analysis.*

1. Sequence spaces

The peculiarities of INT are consequences of continuity principles on spaces of sequences of natural numbers. Working entirely within BISH, in this section we give the basic definitions and facts regarding the two sequence spaces we need in the rest of the chapter. First, we introduce some notation.

Let S be a set. The set of all finite sequences in S, including the empty sequence (), is denoted by S^*. If $a \equiv (a_1,\ldots,a_n)$ is in S^*, then n is called the **length** of a, and is denoted by $|a|$. The set S^* is a monoid under the **concatenation** operation defined by

$$(a_1,\ldots,a_n)(b_1,\ldots,b_m) = (a_1,\ldots,a_n,b_1,\ldots,b_m).$$

If $m \in \mathbb{N}$, and $\alpha \in S^{\mathbb{N}}$ or $\alpha \in S^*$ is of length at least m, then we denote by $\overline{\alpha}(m)$ the finite sequence consisting of the first m terms of α. Note that $\overline{\alpha}(0) = ()$. If $a \in S^*$, and $a = \overline{\alpha}(m)$ for some m, we say that α is an **extension** of a, and that a is a **restriction** of α.

We will be particularly concerned with the sets $\mathbb{N}^*$ and 2^*, where 2 denotes the set $\{0,1\}$. We will identify the natural number k with the element (k) of length one in $\mathbb{N}^*$.

(1.1) Theorem *The sequence space $\mathbb{N}^{\mathbb{N}}$ is a complete, separable metric space under the metric*

$$\rho(a,b) \equiv \inf\{2^{-n} : \bar{a}(n) = \bar{b}(n)\};$$

the subset $2^{\mathbb{N}}$ *of* $\mathbb{N}^{\mathbb{N}}$*, consisting of the binary sequences, is a compact subspace of* $\mathbb{N}^{\mathbb{N}}$*; and there is a uniformly continuous map* $r : \mathbb{N}^{\mathbb{N}} \to \mathbb{N}^{\mathbb{N}}$ *such that* $r \circ r = r$ *and* $r(\mathbb{N}^{\mathbb{N}}) = 2^{\mathbb{N}}$.

Proof. Exercise. □

Note that if $n \in \mathbb{N}$ and $2^{-n-1} < \delta \leq 2^{-n}$, then, relative to the metric ρ of Theorem (1.1),

$$B(a,\delta) = \{b : \bar{a}(n+1) = \bar{b}(n+1)\}.$$

So we can specify open balls in $\mathbb{N}^{\mathbb{N}}$ in terms of a natural number n rather than a positive number δ; as the latter specification is linked to the arbitrary choice of 2 in the definition of the metric ρ, we will prefer the former.

(1.2) Theorem *A function* f *from a subspace* S *of* $\mathbb{N}^{\mathbb{N}}$ *to a metric space* X *is pointwise continuous if and only if for each* $a \in S$ *and each* $\epsilon > 0$*, there exists* $n \in \mathbb{N}$ *such that* $\rho(f(a),f(b)) < \epsilon$ *whenever* $\bar{a}(n) = \bar{b}(n)$*;* f *is uniformly continuous if* n *can be chosen independent of* a.

Proof. Exercise. □

Let π be a map from a metric space T onto a metric space X. We say that π is a **quotient map**, and that X is a **quotient** of T, if a subset S of X is open if and only if its preimage $\pi^{-1}S$ is open. Note that as a function f between metric spaces is pointwise continuous if and only if $f^{-1}S$ is open for each open set S, a quotient map is pointwise continuous. We say that π is a **uniform quotient map**, and that X is a **uniform quotient** of T, if (i) π is uniformly continuous, and (ii) there exists a sequence (δ_n) of positive numbers such that if $x \in X$, then there exists t in T such that $x = \pi(t)$, and $B(x,\delta_n) \subset \pi B(t,2^{-n})$ for all n. The reader may prove that a uniform quotient map is a quotient map.

The main theorem of this section states that each nonvoid separable metric space is a uniform quotient of $\mathbb{N}^{\mathbb{N}}$, and that each nonvoid compact metric space is a uniform quotient of $2^{\mathbb{N}}$. Before proving this, we show how to organize a countable dense subset of a metric space, or a set of ϵ-approximations to a totally bounded space.

(1.3) Lemma *Let X be a nonvoid compact (respectively, separable) metric space, and let $0 < r \leq 1/2$. Then we can construct points $p_\alpha \in X$, indexed by $\alpha \in 2^*$ (respectively, $\mathbb{N}^*$), and a strictly increasing sequence $(m(n))$ in $\mathbb{N}$, such that for each $n \geq 1$,*

(i) $\{p_\alpha : |\alpha| = m(n)\}$ *is an r^n-approximation to X;*

(ii) *if $|\alpha| = m(n)$, then for all β, $\rho(p_\alpha, p_{\alpha\beta}) < r^{n-2}/(1-r)$;*

(iii) *if $|\alpha| = m(n)$ and $\rho(x, p_\alpha) < r^{n-1} - r^{n+1}$, then there exists β such that $|\alpha\beta| = m(n+1)$ and $\rho(x, p_{\alpha\beta}) < r^{n+1}$;*

(iv) *if $|\alpha| = m(n)$ and $|\alpha\beta| < m(n+1)$, then $p_{\alpha\beta} = p_\alpha$.*

Proof. For each $n \geq 0$, let Q_{n+1} be a finite (respectively, countable) r^{n+1}-approximation to X. Let $p_{()}$ be any point of X, and $m(0) \equiv 0$. Having defined $m(n)$ and $\{p_\alpha : |\alpha| = m(n)\}$, for each α of length $m(n)$ partition Q_{n+1} into subsets I_α and J_α, as follows. If $n = 0$, set $I_{()} \equiv Q_1$; otherwise choose I_α and J_α such that

$$q \in I_\alpha \Rightarrow \rho(q, p_\alpha) < r^{n-2},$$
$$q \in J_\alpha \Rightarrow \rho(q, p_\alpha) > r^{n-1}.$$

Choose $m(n+1) > m(n)$ so that for each α of length $m(n)$, the set $\{\beta : |\beta| = m(n+1)-m(n)\}$ can be mapped onto I_α. Set $p_{\alpha\beta} \equiv p_\alpha$ for $|\alpha\beta| < m(n+1)$, and let $\{p_{\alpha\beta} : |\alpha\beta| = m(n+1)\} \equiv I_\alpha$; this is possible by the choice of $m(n+1)$, and completes our inductive construction.

Condition (iv) certainly holds. Condition (i) holds for $n = 1$ as $I_{()} \equiv Q_1$; we shall show that (i) holds for $n = k+1$ if it holds for $n = k$. If (i) holds for $n = k$, then for each q in Q_{k+1} there exists α of length $m(k)$ such that $\rho(q, p_\alpha) < r^k < r^{k-1}$; then $q \in I_\alpha$, and therefore $q = p_{\alpha\beta}$ for some β such that $|\alpha\beta| = m(k+1)$. It follows that $\{p_\alpha : |\alpha| = m(k+1)\}$ is an r^{k+1}-approximation to X; that is, (i) holds for $n = k+1$.

Suppose $n \geq 1$ and $|\alpha| = m(n)$. If $|\alpha\beta| \leq m(n+1)$, then $\rho(p_\alpha, p_{\alpha\beta}) < r^{n-2}$. A simple induction argument shows that if $k \geq 1$ and $|\alpha\beta| \leq m(n+k)$, then

$$\rho(p_\alpha, p_{\alpha\beta}) < r^{n-2}(1 + r + \ldots + r^{k-1});$$

whence (ii) obtains. To prove (iii), suppose that $|\alpha| = m(n)$ and $\rho(x, p_\alpha) < r^{n-1} - r^{n+1}$, and choose q in Q so that $\rho(q, x) < r^{n+1}$. Then $\rho(q, p_\alpha) < r^{n-1}$; whence $q \in I_\alpha$, and (iii) follows by our construction of $p_{\alpha\beta}$ in the case $|\alpha\beta| = m(n+1)$. □

The construction in Lemma (1.3) gives what Brouwer called a

spread representation of the completion of the the space X, by associating with each sequence $\alpha \in \mathbb{N}^{\mathbb{N}}$ the limit of the sequence $(p_{\overline{\alpha}(n)})$.

(1.4) Theorem *If X is a nonvoid complete, totally bounded (respectively, separable) metric space, then there exists a uniform quotient map from $2^{\mathbb{N}}$ (respectively, $\mathbb{N}^{\mathbb{N}}$) onto X.*

Proof. Let D be $\mathbb{N}$ or 2, depending on whether we are in the separable or the totally bounded case. Let $0 < r \leq 1/2$, and construct a family $(p_\alpha)_{\alpha \in D^*}$ of points of X as in (1.3). For each $\alpha \in D^{\mathbb{N}}$, consider the sequence $(p_{\overline{\alpha}(i)})_{i=0}^{\infty}$ of points in X. As $p_{\overline{\alpha}(i)} \in B(p_{\overline{\alpha}(j)}, r^{n-2}/(1-r))$ whenever $i \geq j \geq m(n)$, $(p_{\overline{\alpha}(i)})$ is a Cauchy sequence and so converges to a point $f(\alpha) \in X$; clearly, $\rho(f(\alpha), p_{\overline{\alpha}(m(n))}) \leq r^{n-2}/(1-r)$ for all n.

To show that the function f so defined is uniformly continuous, consider $\alpha \in \mathbb{N}^{\mathbb{N}}$ and $\epsilon > 0$. Choose $n > 0$ so that $r^{n-2}/(1-r) < \epsilon/2$, and suppose that $\overline{\alpha}(m(n)) = \overline{b}(m(n))$. Then

$$f(\alpha), f(b) \in B(p_{\overline{\alpha}(m(n))}, r^{n-2}/(1-r)),$$

so $\rho(f(\alpha), f(b)) < \epsilon$. The uniform continuity of f now follows by (1.2).

To show that f is a uniform quotient map, first note that as $0 < r \leq 1/2$, we have $r^{n-1} - r^{n+1} - r^n > 0$ for all positive integers n. Fix N in $\mathbb{N}^+$, and write

$$\delta \equiv r^{N-1} - r^{N+1} - r^N.$$

Given x in X, use (1.3,i), and then (1.3,iii) inductively, to construct a sequence $\alpha \in D^{\mathbb{N}}$ such that $\rho(x, p_{\overline{\alpha}(m(n))}) < r^n$ for all $n \geq 1$; then $f(\alpha) = x$. Also, if $y \in B(x,\delta)$, then

$$\rho(y, p_{\overline{\alpha}(m(N))}) < \delta + \rho(p_{\overline{\alpha}(m(N))}, x) \leq r^{N-1} - r^{N+1};$$

whence, by repeated application of (1.3,iii), there exists b in $D^{\mathbb{N}}$ such that $y = \lim_{n\to\infty} p_{\overline{b}(n)} = f(b)$ and $\overline{b}(m(N)) = \overline{\alpha}(m(N))$. Therefore $B(x,\delta) \subset fB(\alpha, 2^{-m(N)+1})$. As N is arbitrary and $(m(n))$ is strictly increasing, it follows that f is a uniform quotient map. □

2. Continuous choice

Intuitionistic mathematics diverges from other types of constructive mathematics in its interpretation of the term 'sequence'. A sequence in BISH or RUSS is given by a rule which determines, in advance, how to construct each term of the sequence; such a sequence may be said to be **lawlike** or **predeterminate.** Brouwer, the founder of the intuitionistic school, generalized this notion of a sequence to include the possibility of constructing the terms of a sequence one-by-one, the choice of each term being made freely, or subject only to certain restrictions stipulated in advance. Most manipulations of sequences do not require that they be predeterminate, and can be performed on these more general **free choice sequences.**

Thus, for the intuitionist, a real number (x_n) need not be given by a rule: its terms $x_1, x_2, \ldots$ are simply rational numbers, successively constructed, subject only to the restriction

$$|x_m - x_n| \leq m^{-1} + n^{-1} \quad (m,n \in \mathbb{N}^+).$$

All our work on real numbers in Chapter 1 remains valid under this interpretation.

Let P be a subset of $\mathbb{N}^{\mathbb{N}} \times \mathbb{N}$, and suppose that for each $\alpha \in \mathbb{N}^{\mathbb{N}}$ there exists $n \in \mathbb{N}$ such that $(\alpha,n) \in P$. From a constructive point of view, this means that we have a procedure, applicable to sequences, that computes n for any given α. Now, according to Brouwer, the construction of an element of $\mathbb{N}^{\mathbb{N}}$ is forever incomplete: a generic sequence α is *purely extensional*, in the sense that at any given moment we can know nothing about α other than a finite set of its terms. It follows that for a given sequence α, our procedure must be able to calculate, from some finite initial sequence $\overline{\alpha}(m)$, a natural number n such that $P(\alpha,n)$. If b is another such sequence, and $\overline{\alpha}(m) = \overline{b}(m)$, then our procedure must return the same n for b as it does for α. So this procedure defines a continuous function $f:\mathbb{N}^{\mathbb{N}} \to \mathbb{N}$ such that $(\alpha,f(\alpha)) \in P$ for each $\alpha \in \mathbb{N}^{\mathbb{N}}$. Thus we are led to the following **principle of continuous choice,** which we divide into a continuity part and a choice part.

CC (1) **Any function from $\mathbb{N}^{\mathbb{N}}$ to $\mathbb{N}$ is continuous.**

(2) **If $P \subset \mathbb{N}^{\mathbb{N}} \times \mathbb{N}$, and for each $\alpha \in \mathbb{N}^{\mathbb{N}}$ there exists $n \in \mathbb{N}$ such that $(\alpha,n) \in P$, then there is a function $f:\mathbb{N}^{\mathbb{N}} \to \mathbb{N}$ such that $(\alpha,f(\alpha)) \in P$ for all $\alpha \in \mathbb{N}^{\mathbb{N}}$.**

If P and f are as in CC(2), then we say that f is a **choice function** for P.

Most omniscience principles are demonstrably false under the hypothesis CC; thus we get strong counterexamples, rather than just Brouwerian counterexamples, if we assume CC. For example, we have

(2.1) Proposition LLPO *and* CC *are inconsistent; consequently* LPO *and* WLPO *are inconsistent with* CC.

Proof. Suppose LLPO holds, and define a subset P of $\mathbb{N}^{\mathbb{N}} \times \mathbb{N}$ as follows. Given $a \in \mathbb{N}^{\mathbb{N}}$, define $b \in \mathbb{N}^{\mathbb{N}}$ by setting

$$\begin{aligned} b_n &= 1 \quad \text{if } a_n \neq 0, \text{ and } a_m = 0 \text{ for all } m < n, \\ &= 0 \quad \text{otherwise.} \end{aligned}$$

Then LLPO ensures that either $b_{2n} = 0$ for each n, or else $b_{2n+1} = 0$ for each n. Let $(a,0) \in P$ in the former case, and $(a,1) \in P$ in the latter.

For each i in $\mathbb{N}$, let $\delta^i \in \mathbb{N}^{\mathbb{N}}$ be the sequence whose i^{th} term is 1 and whose other terms are 0. Then $(\delta^{2i},0) \notin P$, and $(\delta^{2i+1},1) \notin P$. Thus, with f as in CC, we have $f(\delta^{2i}) = 1$ and $f(\delta^{2i+1}) = 0$. As (δ^i) converges to $\mathbf{0}$ in the metric on $\mathbb{N}^{\mathbb{N}}$, f is not continuous at $\mathbf{0}$. This contradicts CC. □

Our argument justifying the principle of continuous choice fails if we allow sequences to be given by rules, which may be distinct for equal sequences. If we demand that all sequences be given by rules in the restricted class deemed adequate by Church's thesis, then CC is false, as the following theorem shows.

(2.2) Theorem CC *and* CPF *are inconsistent.*

Proof. Assume that CPF holds. Then for each $a \in \mathbb{N}^{\mathbb{N}}$ there exists $n \in \mathbb{N}$ such that $a = \varphi_n$. Let

$$P \equiv \{(a,n) \in \mathbb{N}^{\mathbb{N}} \times \mathbb{N} : a = \varphi_n\}.$$

Choose k so that $\mathbf{0} = \varphi_k$. Suppose there were a pointwise continuous function f from $\mathbb{N}^{\mathbb{N}}$ to $\mathbb{N}$ such that $(a,f(a)) \in P$ for all a. Then each nonzero $a \in \mathbb{N}^{\mathbb{N}}$ sufficiently close to $\mathbf{0}$ would satisfy $f(a) = f(\mathbf{0})$, and therefore $a = 0$, which is certainly false. □

As CC(1) is consistent with CPF (see [Beeson, Chap. XVI, §2]), the inconsistency between CC and CPF stems from the strong choice principle embodied in CC(2).

The next theorem shows that CC(1) may be strengthened by replacing the target space $\mathbb{N}$ by an arbitrary metric space.

(2.3) Theorem *If CC holds, then any function from* $\mathbb{N}^{\mathbb{N}}$ *or* $2^{\mathbb{N}}$ *to a metric space is pointwise continuous.*

Proof. Let f be a function from $\mathbb{N}^{\mathbb{N}}$ to a metric space, and let $a \in \mathbb{N}^{\mathbb{N}}$. Given $\epsilon > 0$, define a subset P of $\mathbb{N}^{\mathbb{N}} \times \mathbb{N}$ by

$$P \equiv \{(b,1) : \rho(f(a),f(b)) < \epsilon\} \cup \{(b,0) : \rho(f(a),f(b)) > 0\}.$$

Then for each b in $\mathbb{N}^{\mathbb{N}}$ there exists n in $\{0,1\}$ such that $(b,n) \in P$. Let g be a continuous choice function for P. As $g(a) = 1$, there exists $\delta > 0$ such that $g(B(a,\delta)) \subset \{1\}$; so if $b \in B(a,\delta)$, then $\rho(f(a),f(b)) < \epsilon$.

Let $r:\mathbb{N}^{\mathbb{N}} \to 2^{\mathbb{N}}$ be the retraction introduced in Theorem (1.1). If g is a function from $2^{\mathbb{N}}$ to a metric space Y, then $gr:\mathbb{N}^{\mathbb{N}} \to Y$ is continuous, by the first part of the theorem. Therefore gr restricted to $2^{\mathbb{N}}$ is continuous; but gr restricted to $2^{\mathbb{N}}$ is g. □

(2.4) Corollary *Let* X *be a nonvoid, complete, separable metric space, and* Y *a metric space. If CC holds, then each function* $g:X \to Y$ *is pointwise continuous.*

Proof. By (1.4), there is a (uniform) quotient map f from $\mathbb{N}^{\mathbb{N}}$ onto X. Then gf is a map from $\mathbb{N}^{\mathbb{N}}$ to Y, and so is pointwise continuous, by (2.3). If S is an open set in Y, then $(gf)^{-1}S$ is open in X. But $(gf)^{-1}S = f^{-1}(g^{-1}S)$; so $g^{-1}S$ is open, as f is a quotient map. □

Taking $X \equiv \mathbb{R}$ in Corollary (2.4), we get Brouwer's famous theorem that every function from the reals to the reals is continuous:

(2.5) Corollary *If CC holds, then every map from* $\mathbb{R}$ *to* $\mathbb{R}$ *is pointwise continuous.* □

We conclude this section with an application to linear operators on normed linear spaces.

(2.6) Theorem *Let* X *be a complete, separable normed linear space and* (u_n) *a sequence of linear maps from* X *to a normed linear space such that for each unit vector* x *of* X,

$$\varphi(x) \equiv \sup\{\|u_n(x)\| : n \in \mathbb{N}\}$$

exists. If CC holds, then there exists $c > 0$ *such that* $\|u_n(x)\| \le c$ *for all* $n \in \mathbb{N}$ *and all unit vectors* x *of* X.

Proof. First note that as each u_n is homogenous, so is φ. By (2.4), the map $\varphi:X \to \mathbb{R}$ is continuous at 0. Thus there exists $c > 0$ such that

$\varphi(z) \leq 1$ for all z in X with $\|z\| \leq c^{-1}$. For any unit vector x in X we have $c^{-1}\varphi(x) = \varphi(c^{-1}x) \leq 1$; so that $\|u_n(x)\| \leq \varphi(x) \leq c$ for each n. □

(2.7) Corollary *If CC holds, then each linear map from a complete, separable normed linear space into a normed linear space is bounded.*

Proof. Take all the u_n in (2.6) to be the same. □

Corollary (2.7) is false without CC, but the classical counterexamples are rather wild. For example, the set C of continuous functions on $[0,1]$ is dense in the complete, separable normed linear space $L^1[0,1]$, and the map taking f to $f(0)$ is unbounded on C in the L^1 norm. Zorn's lemma allows us (classically) to extend this map to a linear map on L^1 that is unbounded.

3. Uniform continuity

To deal with the uniform continuity of functions on a compact metric space, we introduce some more notions due to Brouwer.

A detachable subset σ of $\mathbb{N}^*$ is called a **fan** if it is closed under restriction, and for each $a \in \sigma$ the set $\{n \in \mathbb{N} : an \in \sigma\}$ is nonempty and finite. For example, the set 2^* is a fan, called the **complete binary fan**; the following pictorial representation of 2^* indicates where the name 'fan' comes from:

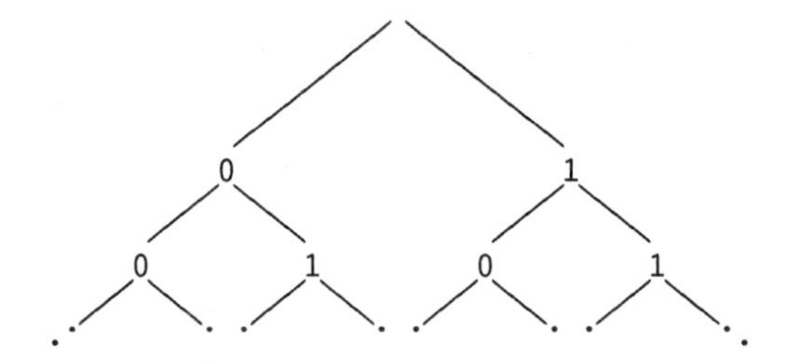

A **path** in a fan σ is a sequence α, finite or infinite, such that $\overline{\alpha}(n) \in \sigma$ for each applicable n. We say that a path α is **blocked** by a subset B if some restriction of α is in B; if no restriction of α is in B we say that α **misses** B. A subset B of a fan σ is called a **bar** for σ if each infinite path of σ is blocked by B; a bar B for σ is **uniform** if there exists n in $\mathbb{N}$ such that each path of length n is blocked by B.

The following principle, the **fan theorem**, is used heavily in intuitionistic mathematics.

FT Every detachable bar of a fan is uniform.

In its classical contrapositive form, the fan theorem is known as **König's lemma**: if for every n there exists a path of length n that misses B, then there exists an infinite path that misses B (see **Dummett**, pp. 68–71). It is easy to construct Brouwerian counterexamples to König's lemma.

Given a detachable bar B of a fan σ, one can construct a detachable bar of 2^* that is uniform if and only if B is uniform (Problem 5). Thus the fan theorem is equivalent to its restriction to the fan 2^*.

Attempts to prove the fan theorem constructively rely on an analysis of how we could know that a subset is a bar. Let σ be a fan and B a subset of σ. Consider the set A of elements α of σ such that any infinite path extending α is blocked by B. To establish that B is a bar, we must show that the empty sequence is in A. Let B_0 consist of those elements of σ that have restrictions in B; then, clearly, $B_0 \subset A$. Having constructed B_n, let B_{n+1} consist of those elements α of σ such that $\alpha\beta \in B_n$ for each β of length 1. It is easily seen by induction that $B_n \subset A$ for all n. If, as was the case with Brouwer, one can be convinced, somehow, that the only way to show that a finite sequence is in A is to show that it is in B_n for some n, then one can establish the fan theorem. For it then follows, from the intuitionistic interpretation of implication, that $A = \cup B_n$; so that if B is a bar, and we choose N such that the empty sequence is in B_N, then every path of length N is blocked by B.

Note that we do not require that B be detachable in the above argument, although possibly the required leap of faith is easier in that case.

Brouwer extended this kind of argument from a fan to $\mathbb{N}^*$. In that setting, we cannot stop at the construction of sets B_n: for we might have an element α of σ such that $\alpha\beta \in \cup B_n$ whenever β is of length 1, but $\alpha \notin \cup B_n$; such elements form another set B_ω. Continuing in this way we get a well-ordered collection of subsets B_θ whose union is A. Rather than consider well-orderings, we can rephrase the argument in terms of an induction principle called **bar-induction**: If $B \subset B'$, and $\alpha \in B'$ whenever $\alpha\beta \in B'$ for each β of length 1, then $A \subset B'$. Although bar-induction actually admits Brouwerian counterexamples in its most general form, it appears to cause no constructive problems if we add the condition that B be detachable (**Dummett**, 74–78).

As the fan theorem is a fundamental part of intuitionistic mathematics, we shall spend some time exploring its consequences. First, we show that although the fan theorem may be plausible from Brouwer's point of view, it need not hold in a development of constructive mathematics which eschews the notion of a choice sequence; in particular, it conflicts with Church's thesis.

(3.1) Proposition *If CPF holds, then there is a detachable bar for* 2^* *that is missed by arbitrarily long finite paths.*

Proof. For each integer $r \geq 2$ and each prime number p, let $r[p]$ be the exponent of p in the prime factorisation of r. Assuming CPF, and using the notation of Chapter 3, define a partial function $w:\mathbb{N}^2 \to \{0,1\}$ as follows: if $r \in D_{r[3]}(k)\backslash D_{r[3]}(k-1)$, and $r \notin D_{r[2]}(k)$, then $w(r,k) \equiv 0$; if $r \in D_{r[2]}(k)\backslash D_{r[2]}(k-1)$, and $r \notin D_{r[3]}(k)$, then $w(r,k) \equiv 1$. Note that the domain of w is detachable. Let B be the detachable subset of 2^* consisting of all finite binary sequences $(a_1, \ldots, a_n)$ such that $w(r,k) = a_r$ for some $r < n$ and some $k < n-r$. Note that any extension of an element of B is in B.

We first prove that B is a bar. Fix an infinite path $a = (a_1, a_2, \ldots)$ in 2^*, choose p and q such that

$$\varphi_p(n) = \begin{cases} 0 & \text{if } a_n = 0, \\ \text{undefined} & \text{otherwise,} \end{cases}$$

and

$$\varphi_q(n) = \begin{cases} 1 & \text{if } a_n = 1, \\ \text{undefined} & \text{otherwise,} \end{cases}$$

and let $r \equiv 2^q 3^p$. Then $w(r,k) = a_r$ for some k: for example, if $a_r = 0$, then $\varphi_p(r) = 0$ and $\varphi_q(r)$ is undefined, so $r \in D_{r[3]}(k)\backslash D_{r[3]}(k-1)$ for some k, and $r \notin D_{r[2]}(k)$. With this choice of k, it follows that $\overline{a}(k+r+1) \in B$. Hence B is a bar.

We now prove that for each positive integer n there is a path of length n missing B. To this end, for $r = 1, \ldots, n$ define

$$a_r = \begin{cases} 1 & \text{if } r < n \text{ and } w(r,k) = 0 \text{ for some } k < n-r, \\ 0 & \text{otherwise.} \end{cases}$$

If $(a_1, \ldots, a_n) \in B$, then there exist $r < n$ and $k < n-r$ such that

$w(r,k) = a_r$. If $a_r = 0$, then $w(r,k) = a_r = 0$, and the definition of a_r leads to the contradiction $a_r = 1$. Thus $a_r = 1$, and so $w(r,k) = a_r = 1$. Again using the definition of a_r, we see that $w(r,k') = 0$ for some $k' < n-r$. But we cannot have both $w(r,k') = 0$ and $w(r,k) = 1$. Thus, in fact, $(a_1,\ldots,a_n) \notin B$. □

The fan theorem and CC can be used to derive the **uniform continuity principle:**

UC Every pointwise continuous function from $2^{\mathbb{N}}$ to $\mathbb{N}$ is uniformly continuous.

In fact, as we shall see, under the hypothesis CC, the fan theorem and the uniform continuity principle are equivalent within BISH.

(3.2) Theorem *If CC holds, and each detachable bar for 2^* is uniform, then every function from $2^{\mathbb{N}}$ to $\mathbb{N}$ is uniformly continuous.*

Proof. Consider a function $f:2^{\mathbb{N}} \to \mathbb{N}$. By CC, there is a continuous function $g:2^{\mathbb{N}} \to \mathbb{N}$ such that if $\overline{a}(g(a)) = \overline{b}(g(a))$, then $f(a) = f(b)$. For each finite sequence α, let $\alpha^* = \alpha000\cdots$. Define a detachable subset B of $2^{\mathbb{N}}$ by setting $B \equiv \{\alpha : g(\alpha^*) \leq |\alpha|\}$. To show that B is a bar, consider any a in $2^{\mathbb{N}}$, and choose $n \geq g(a)$ such that if $\overline{a}(n) = \overline{b}(n)$, then $g(a) = g(b)$; then

$$g(\overline{a}(n)^*) = g(a) \leq n = |\overline{a}(n)|,$$

and so $\overline{a}(n) \in B$. If each detachable bar is uniform, then there exists N such that for each a in $2^{\mathbb{N}}$, $g(\overline{a}(n)^*) \leq n$ for some $n \leq N$. If $\overline{a}(N) = \overline{b}(N)$, and $g(\overline{a}(n)^*) \leq n \leq N$, then $f(a) = f(\overline{a}(n)^*) = f(b)$; whence f is uniformly continuous. □

Detachable bars give rise to continuous functions which are uniformly continuous precisely when the bar is uniform.

(3.3) Theorem *Let B be a detachable bar for 2^*. Then there is a unique pointwise continuous function $g:2^{\mathbb{N}} \to \mathbb{N}$ such that $\overline{a}(g(a)) \in B$, and $\overline{a}(m) \notin B$ for each $m < g(a)$; the bar B is uniform if and only if g is uniformly continuous.*

Proof. It is easy to see that the unique pointwise continuous function g fulfilling our requirements is defined by setting

$$g(a) \equiv \min\{n : \overline{a}(n) \in B\},$$

and that if B is uniform, then g is uniformly continuous. Conversely, suppose that g is uniformly continuous, and choose N such that if $\overline{a}(N) = \overline{b}(N)$, then $g(a) = g(b)$. Since $\{\overline{a}(N) : a \in 2^{\mathbb{N}}\}$ is finite, the range of g is a finite subset of $\mathbb{N}$; whence $\max\{g(a) : a \in 2^{\mathbb{N}}\}$ exists, and B is therefore uniform. □

(3.4) Corollary *If* UC *holds, then each detachable bar for* 2^* *is uniform.*

Proof. Let B be a detachable bar for 2^*, and let g be the pointwise continuous function g, corresponding to B, supplied by (3.3). Then g is uniformly continuous, by UC; so B is uniform, by (3.3). □

For the remainder of this chapter, we shall assume both the principle of continuous choice and the fan theorem. Thus, in effect, we shall be working in the full context of INT.

The following theorem is a very general version of the Heine-Borel theorem of elementary analysis. Note that this theorem, as stated, is false in classical mathematics, since the latter admits noncompact metric spaces that are images, but not continuous images, of $2^{\mathbb{N}}$.

(3.5) Theorem *If the metric space* X *is an image of* $2^{\mathbb{N}}$, *then every open cover of* X *has a finitely enumerable subcover.*

Proof. Let g be a mapping of $2^{\mathbb{N}}$ onto X, and $(U_i)_{i\in I}$ an open cover of X. By (2.3), the map g is pointwise continuous; whence X is separable. Let $(x_n)_{n=0}^{\infty}$ be a dense sequence in X, and

$$P \equiv \{(a,m,n) \in 2^{\mathbb{N}} \times \mathbb{N}^2 : g(a) \in B(x_m, 2^{-n}) \subset U_i \text{ for some } i \text{ in } I\}.$$

Then for each a in $2^{\mathbb{N}}$ there exist m and n such that $(a,m,n) \in P$; so, by CC and UC, there exists a uniformly continuous function $f : 2^{\mathbb{N}} \to \mathbb{N}^2$ such that $(a, f(a)) \in P$ for all a in $2^{\mathbb{N}}$. Choose v such that if $a,b \in 2^{\mathbb{N}}$ and $\overline{a}(v) = \overline{b}(v)$, then $f(a) = f(b)$. Since $\{\overline{a}(v) : a \in 2^{\mathbb{N}}\}$ is a finite set, the set $f(2^{\mathbb{N}})$ is finite. For each (m,n) in $f(2^{\mathbb{N}})$, choose $i(m,n)$ in I such that $B(x_m, 2^{-n}) \subset U_{i(m,n)}$. Then $(U_{i(m,n)})_{(m,n)\in f(2^{\mathbb{N}})}$ is a finitely enumerable subcover of (U_i). □

We can now prove the **uniform continuity theorem.**

(3.6) Theorem *Every mapping of a nonvoid compact metric space into a metric space is uniformly continuous.*

Proof. Let f be a mapping of a nonvoid compact metric space X into a metric space. By (2.4), f is pointwise continuous. By (1.4) and (3.5), every open cover of X has a finitely enumerable subcover. Given $\epsilon > 0$, consider

$$I \equiv \{(x,r) : r > 0 \text{ and } fB(x,2r) \subset B(f(x),\epsilon/2)\}.$$

As $\{B(x,r) : (x,r) \in I\}$ is an open cover of X, there is a finitely enumerable subset J of I such that $\{B(x,r) : (x,r) \in J\}$ is a cover of X. Let

$$\delta \equiv \frac{1}{2} \min\{r : (x,r) \in J \text{ for some } x \text{ in } X\}.$$

If $\rho(x_1,x_2) < \delta$, choose (x,r) in J such that $\rho(x,x_1) < r$. Then $\rho(x,x_2) < 2r$; so that $\rho(f(x),f(x_i)) < \epsilon/2$ for each i, and therefore $\rho(f(x_1),f(x_2)) < \epsilon$. □

(3.7) Corollary *If f is a function from a nonvoid compact metric space into $\mathbb{R}^+$, then* $\inf f$ *exists and is positive.*

Proof. The function $1/f$ is uniformly continuous, by (3.6); whence, by (4.5) of Chapter 2, it has a supremum M. Clearly, $M > 0$ and $\inf f = 1/M > 0$. □

The following lemma will enable us to prove that a compact interval of the real line has no nontrivial detachable subsets.

(3.8) Lemma *Every integer-valued function on a compact interval is constant.*

Proof. Let f be an integer-valued function on the compact interval $I \equiv [a,b]$. By (3.6), there exists $\delta > 0$ such that if $|x-y| < \delta$, then $|f(x)-f(y)| < 1/2$, and therefore $f(x) = f(y)$. Given $x \in I$, choose points $a = p_1 \leq p_2 \leq \dots \leq p_n = x$ of I such that $p_{k+1} - p_k < \delta$ for $k = 1,\dots,n-1$. Then $f(p_k) = f(p_{k+1})$ for each such k, and so $f(x) = f(a)$. □

(3.9) Proposition *If S is a nonvoid detachable subset of a compact interval I, then $S = I$.*

Proof. Define a function $f:I \to \{0,1\}$ by setting $f(x) \equiv 1$ if $x \in S$, and $f(x) \equiv 0$ if $x \notin S$. Then f is constant, by (3.8); so that, as S is nonvoid $f(x) = 1$ for all x in I, and therefore $S = I$. □

4. The creating subject and Markov's principle.

a simple and clear example, which I gave now and then in courses and lectures since 1927 (L.E.J. Brouwer)

The example to which Brouwer was referring, a counterexample to Markov's principle, uses the sequential thought process of an idealized mathematician, or **creating subject**, to construct a special sequence, and thereby illuminates Brouwer's view of the nature of choice sequences. Another aspect of the example is that it makes essential use of the intuitionistic interpretation of $P \Rightarrow Q$: namely, there exists a finite routine for converting any proof of P into a proof of Q; in particular, to prove $\neg P$ it suffices to show that it would be absurd (for the creating subject) to prove P.

For the creating subject, time is divided into discrete stages, during each of which he may test various propositions, attempt to construct proofs, and so on. In particular, we can determine whether or not at stage n the creating subject has a proof of a particular mathematical assertion P.

Given any mathematical assertion P, define a sequence of rational numbers (x_n) as follows. If, at stage n, the creating subject has neither a proof of P nor a proof of $\neg P$, then $x_n \equiv 0$. If between stage $n-1$ and stage n the creating subject has obtained a proof of P, then $x_m \equiv 2^{-n}$ for each $m \geq n$. If between stage $n-1$ and stage n the creating subject has obtained a proof of $\neg P$, then $x_m \equiv -2^{-n}$ for each $m \geq n$.

As (x_n) is a Cauchy sequence, it converges to a real number r. If we could prove that $r = 0$, then we could show that it would be absurd for the creating subject to prove P, and that it would be absurd for the creating subject to prove $\neg P$. Thus we would have proved $\neg P$ and $\neg\neg P$, which is absurd. So r cannot be 0. It follows from Markov's principle that $r \neq 0$. But this means that we can either demonstrate P or demonstrate $\neg P$. Thus Markov's principle entails the law of excluded middle.

Because it introduces temporal considerations into mathematics, the theory of the creating subject is unacceptable to most mathematicians, even among the constructive fraternity. However the mathematical consequences, if not the substance, of that theory can be obtained without reference to time, by a simple postulate. Of course, such a postulate may be just as controversial as the theory it is designed to avoid.

The creating subject allows us to define, for a given proposition P, a binary sequence (a_n) as follows:

$$a_n = 1 \quad \text{if the creating subject has a proof of } P \text{ at stage } n,$$
$$= 0 \quad \text{otherwise.}$$

If the construction of these sequences is the only use we make of the creating subject, then references to the creating subject can be avoided by postulating the principle known as **Kripke's Schema:**

> *For each proposition P there exists an increasing binary sequence (a_n) such that P holds if and only if $a_n = 1$ for some n.*

Thus Kripke's schema says that every proposition is simply existential, in the sense of Chapter 1. Brouwer's argument above shows that Kripke's schema, together with Markov's principle, implies the law of excluded middle; as the latter is contradictory in INT, Markov's principle is refutable in INT + Kripke's schema.

PROBLEMS

1. Show that the two conditions in the definition of a uniform quotient map are independent.

2. Show that CC is equivalent to the following principle:

   If to each x in $\mathbb{N}^{\mathbb{N}}$ there corresponds n in $\mathbb{N}$ with the property $P(x,n)$, then there exist a mapping $f:\mathbb{N}^{\mathbb{N}} \to \mathbb{N}$ and a mapping r of $\mathbb{N}^{\mathbb{N}}$ onto a detachable subset of $\mathbb{N}^*$, such that, for all x and y in $\mathbb{N}^{\mathbb{N}}$,

   (i) $P(x,f(x))$;

   (ii) $r(x)$ is a restriction of x;

   (iii) if $r(x)$ is a restriction of $r(y)$, then $f(x) = f(y)$.

3. Prove within BISH that the following extension of CC(1) is not consistent with CPF: if f is a map of $\mathbb{N}^{\mathbb{N}}$ into $\mathbb{N}$, then there exists a map $g:\mathbb{N}^{\mathbb{N}} \to \mathbb{N}^{\mathbb{N}}$ such that for all a in $\mathbb{N}^{\mathbb{N}}$,

   (i) $f(a) = f(b)$ whenever $b \in \mathbb{N}^{\mathbb{N}}$ and $\overline{a}(g(a)) = \overline{b}(g(a))$, and

   (ii) if $n < g(a)$, then there exists b in $\mathbb{N}^{\mathbb{N}}$ such that

$$\overline{a}(n) = \overline{b}(n) \text{ and } f(a) \neq f(b).$$

(Consider the map $f: \mathbb{N}^{\mathbb{N}} \to \mathbb{N}^{\mathbb{N}}$ defined by

$$\begin{aligned} f(a) &= 0 \quad \text{if } a_0 \in D_{a_0}(a_1) \backslash D_{a_0}(a_1-1), \\ &= 1 \quad \text{otherwise.} \end{aligned}$$

Suppose g exists with the properties (i) and (ii), and derive a contradiction to Lemma (1.2) of Chapter 3)

4. Prove the following intuitionistic version of **Lindelöf's theorem:** If X is a separable metric space, then every open cover of X has a countable subcover.

5. Given a fan σ, construct a function $\varphi:\sigma \to 2^*$ such that a is a restriction of b if and only if $\varphi(a)$ is a restriction of $\varphi(b)$. Show that $\varphi(\sigma)$ is a detachable subset of 2^*. If B is a detachable bar of σ, extend $\varphi(B)$ to a detachable bar of $\varphi(\sigma)$ that is uniform if and only if B is uniform.

6. Give a Brouwerian counterexample to König's lemma.

7. Let S be a detachable subset of $\mathbb{N}$. Let B be the subset of $\mathbb{N}^*$ consisting of $\{(n) : n \notin S\} \cup \{() : S \text{ is nonempty}\}$. Show that B is a bar for $\mathbb{N}^*$. Let $B' = B \cup \{() : S \text{ is empty}\}$. Use B and B' to give a Brouwerian counterexample to the full form of bar-induction.

8. Give an intuitionistic proof of the **general fan theorem:** every bar (not necessarily detachable) for a fan is uniform.

9. Give an intuitionistic proof of Theorem (3.6), without invoking Theorem (3.5), directly from the uniform continuity principle.

10. Prove intuitionistically that any image of $2^{\mathbb{N}}$ is totally bounded.

11. Prove intuitionistically that if X is a compact metric space, and (f_n) is a sequence of real-valued functions on X that converges pointwise to a function f on X, then (f_n) converges to f uniformly on X.

12. Show that if Kripke's schema and LLPO hold, then so does the **weak law of excluded middle:** for each statement P, either $\neg P$ or $\neg\neg P$.

NOTES

Good references for traditional intuitionism are **Dummett** and **Brouwer**. Kleene and Vesley's *The Foundations of Intuitionistic Mathematics*, (North-Holland, 1965) contains a wealth of detail about the formal development of intuitionistic mathematics, but is hard to read.

The principle of continuous choice is equivalent to the following principle, which is called $CP_{\exists n}$ on page 81 of **Dummett**, and $CONT_0$ on page 1006 of Troelstra's article *Aspects of constructive mathematics*, in *Handbook of Mathematical Logic* (J. Barwise, ed., North-Holland 1977):

If for all $\alpha \in \mathbb{N}$, there exists $n \in \mathbb{N}$ such that $P(\alpha,n)$, then there is $g:\mathbb{N}^* \to \mathbb{N}$ such that

(i) $\{\alpha \in \mathbb{N}^* : g(\alpha) > 0\}$ is a bar;

(ii) if $g(\alpha) > 0$, then $g(\alpha\beta) = g(\alpha)$ for all $\beta \in \mathbb{N}^*$;

(iii) for each $\alpha \in \mathbb{N}^{\mathbb{N}}$, if $g(\overline{\alpha}(n)) > 0$, then $P(\alpha,g(\overline{\alpha}(n))-1)$.

The Brouwerian counterexample to bar induction is due to S. C. Kleene; the form we use in Problem 7 comes from **Brouwer** (note 39, page 102).

The quote at the beginning of Section 4 comes from Brouwer's paper *Essentieel negatieve eigenschappen* (Indag. Math. 10 (1948), 322–323), and is available on page 478 of his collected works, edited by A. Heyting, North-Holland, 1975.

An interesting discussion of the theory of the creating subject is found on pages 335–359 of **Dummett**.

Chapter 6. Contrasting Varieties

In which our three varieties of constructive mathematics are compared and contrasted by a study of the proposition that a uniformly continuous map of [0,1] *into the positive real line has positive infimum.*

1. The Three Varieties

The time has come to draw breath and take a comparative look at our three varieties of constructive mathematics. Having done so, we shall undertake a special case-study which will highlight some of the non-philosophical features that distinguish INT and RUSS from BISH.

We begin with BISH, which grows out of the fundamental idea of interpreting 'existence' in terms of an intuitive notion of 'computability'. A most significant feature of BISH is that it is consistent with classical mathematics: every proof in BISH of a statement T is also a classical proof of T. Thus BISH is a generalization of classical mathematics in the same sense that group theory is a generalization of abelian group theory. We pass from BISH to classical mathematics by adopting the law of excluded middle.

BISH is widely accepted as the common core of INT, RUSS, and any genuinely constructive development of mathematics other than a strictly finitistic one. It follows that, within BISH alone, one cannot disprove anything that is provable in INT or RUSS. In particular, we cannot refute, in BISH, the proposition

(1.1) *Every map* $f:[0,1] \to \mathbb{R}$ *is pointwise continuous,*

which is provable in both INT and RUSS. On the other hand, a proof of (1.1) within BISH would also be a classical proof of (1.1), which is impossible. For the same reason, we cannot prove within BISH that the limited principle of omniscience is false; so the Brouwerian counterexamples on which we rely for guidance about the constructive status of many classical theorems are not counterexamples in the usual sense.

We cannot prove or disprove, within BISH alone, the proposition

(1.2) *Every pointwise continuous map* $f:[0,1] \to \mathbb{R}$ *is uniformly continuous,*

which is true in INT, but false in RUSS and false classically. Because of this, BISH is not concerned with proving (1.1) or (1.2), but restricts attention to those maps $f:[0,1] \to \mathbb{R}$ that are known to be uniformly continuous.

To pass from BISH to RUSS, we add postulates, such as CPF, which are consequences of Church's Thesis that all (computable) sequences of natural numbers are recursive. Not only can we then disprove the limited principle of omniscience, thereby turning many Brouwerian counterexamples into genuine counterexamples, but also, with the aid of Markov's Principle, we can establish (1.1). Moreover, by arguments based on the Specker sequence, we can construct remarkable examples like that of an unbounded, pointwise continuous map of $[0,1]$ into $\mathbb{R}$.

From a classical point of view, the recursive real line is countable, so these examples may not seem so remarkable. However, the (recursive) real line is not countable *within RUSS*, as Cantor's theorem holds in BISH ((1.4) of Chapter 2).

The initial development of INT rested heavily on Brouwer's subjectivist philosophy and consequent analysis of the notion of a sequence. This analysis leads to the principle of continuous choice and the fan theorem (or, in Brouwer's original development, the principle of bar induction, from which the fan theorem follows). Taken with the standard notions and techniques of BISH, the principle of continuous choice alone enables us to prove (1.1) and disprove the limited principle of omniscience; the additional hypothesis of the fan theorem leads to a proof of (1.2).

INT and RUSS are formally inconsistent because, for example, (1.2) is true in INT and false in RUSS. From a semantic point of view, this inconsistency arises from the intuitionist's extension of the notion of a sequence, far beyond the limits of recursiveness, to allow free choice in the construction of any term.

RUSS is also formally inconsistent with classical mathematics, since it leads to classically false results like (1.1). The analogue of (1.1) in recursive mathematics, on the other hand, is the classically true proposition

(1.3) *Every recursive map from the set X of recursive real numbers in $[0,1]$ into the recursive real line is pointwise continuous on X,*

and the proof of (1.1) in RUSS can be translated into a classical proof of (1.3). In fact, (1.1) says more in RUSS than is explicit in the statement (1.3), for it requires us to find, for each x in X, a *recursive* sequence (n_k) of positive integers such that $|f(x) - f(y)| < 1/k$ whenever $y \in X$ and $|x - y| < 1/n_k$.

INT, too, is formally inconsistent with classical mathematics; it diverges from classical mathematics at the introduction of the principle of continuous choice, which is incompatible with the law of excluded middle. The fan theorem, on the other hand, is classically valid. As we explained in Chapter 5, the principle of continuous choice arises from the intuitionistic notion of a choice sequence, which contrasts sharply with the classical notion of a predeterminate sequence.

2. Positive-valued Continuous Functions

After the brief discussion in Section 1, it is instructive to consider in some detail the proposition

Every uniformly continuous map from $[0,1]$ to $\mathbb{R}^+$ has a positive infimum

within BISH. We shall show that this proposition is equivalent to the fan theorem, and also give the long-postponed example in RUSS of a uniformly continuous map $f:[0,1] \to \mathbb{R}^+$ whose infimum is zero.

Throughout this section, C will denote the fan of all finite sequences in $\{-1,1\}$, and the first index of all sequences will be 1. With each element $a \equiv (a_1,\ldots,a_n)$ of C we associate a point (x_a, y_a) in the euclidean plane $\mathbb{R}^2$ by setting

$$x_a \equiv 1/2 + \Sigma_{i=1}^{n} a_i 3^{-i},$$

$$y_a \equiv 1/2 - \Sigma_{i=1}^{n} 3^{-i} = 3^{-n}/2.$$

In particular, we also write $x_{()} = y_{()} \equiv 1/2$. For each subset S of C let

$$P(S) \equiv \{(x_a, y_a) : a \in S\}.$$

Since each point of $P(C)$ is isolated in $P(C)$, any finite subset of $P(C)$ is detachable from $P(C)$. Also, if $a, b \in C$ and $a \neq b$, then

$$\|(x_a, y_a) - (x_b, y_b)\| \geq 3^{-|a|-1}\sqrt{2},$$
$$|x_a - x_b| \geq 3^{-|a|-1},$$

and equality obtains in each case if $a = \overline{b}(|b|-1)$.

Before arriving at a substantial result, we need several lemmas.

(2.1) Lemma *Let S be a subset of C. If S is detachable from C and closed under restriction, then P(S) is totally bounded; if P(S) is totally bounded, then S is detachable from C.*

Proof. If S is detachable from C and closed under restriction, then for each n in $\mathbb{N}^+$,

$$S_n \equiv \{a \in S : |a| \leq n\}$$

is finite. Hence $P(S_n)$ is finite. It is also a 3^{-n}-approximation to $P(S)$: for if $a \in S$, then either $a \in S_n$ or $|a| > n$; in the latter case, $\overline{a}(n) \in S_n$, and

$$\|(x_a, y_a) - (x_{\overline{a}(n)}, y_{\overline{a}(n)})\| \leq ((\Sigma_{i=n+1}^{|a|} 3^{-i})^2 + (3^{-n}/2 - 3^{-|a|}/2)^2)^{1/2}$$
$$\leq ((3^{-n}/2)^2 + (3^{-n}/2)^2)^{1/2} < 3^{-n}.$$

Since n is arbitrary, $P(S)$ is totally bounded.

Conversely, assuming that $P(S)$ is totally bounded, and given any element a of C, construct a finite $3^{-|a|-1}$-approximation F to $P(S)$. Since, for any b in S with $b \neq a$, the distance from (x_a, y_a) to (x_b, y_b) is at least $3^{-|a|-1}\sqrt{2}$, it readily follows that $a \in S$ if and only if $(x_a, y_a) \in F$. But F, being a finite subset of $P(C)$, is detachable; hence it is decidable whether or not $a \in S$. □

(2.2) Lemma *Let $x \in \mathbb{R}$, $a \in C$, and $n \in \mathbb{N}^+$, and suppose that either $a_n = 1$ and $x_{\overline{a}(n-1)} > x$, or $a_n = -1$ and $x_{\overline{a}(n-1)} < x$. Then $|x_{\overline{a}(m)} - x| > 3^{-n}/2$ for all $m \geq n$.*

Proof. Consider, for example, the case where $a_n = 1$ and $x_{\overline{a}(n-1)} > x$. If

$m \geq n$, we have

$$\begin{aligned} x_{\overline{\alpha}(m)} - x &= (x_{\overline{\alpha}(n-1)} - x) + 3^{-n} + \Sigma_{i=n+1}^{m} a_i 3^{-i} \\ &> 3^{-n} - \Sigma_{i=n+1}^{\infty} 3^{-i} = 3^{-n}/2. \quad \square \end{aligned}$$

(2.3) Lemma *For each x in $\mathbb{R}$ there is an infinite path a in C such that if S is a subset of C that is closed under restriction, and $\overline{a}(m) \notin S$ for some m in $\mathbb{N}^+$, then for each b in S,*

(2.3.1) $$\|(x,0) - (x_b,y_b)\| \geq 3^{-m}/2.$$

Proof. Given x in $\mathbb{R}$, construct an infinite path $a \equiv (a_n)$ in C such that

$$a_n = 1 \Rightarrow x_{\overline{a}(n-1)} < x,$$

and

$$a_n = -1 \Rightarrow x_{\overline{a}(n-1)} > x - 3^{-n}/5.$$

Let S be a subset of C that is closed under restriction, and suppose that $\overline{a}(m) \notin S$. Then either $|x - x_{\overline{a}(k-1)}| < 3^{-k}/4$ for some $k \leq m$, or else $|x - x_{\overline{a}(n-1)}| > 3^{-n}/5$ for all $n \leq m$. In the first case, for any b in S we have

$$\|(x,0) - (x_b,y_b)\| \geq |y_b| = 3^{-k+1}/2 > 3^{-m}/2$$

if $b = \overline{a}(k-1)$, and

$$\begin{aligned} \|(x,0) - (x_b,y_b)\| &\geq |x - x_b| \\ &\geq |x_{\overline{a}(k-1)} - x_b| - |x - x_{\overline{a}(k-1)}| \\ &\geq 3^{-k} - 3^{-k}/4 > 3^{-k}/2 \geq 3^{-m}/2 \end{aligned}$$

if $b \neq \overline{a}(k-1)$; whence (2.3.1) holds.

In the second case, if $1 \leq n \leq m$, we have $x_{\overline{a}(n-1)} < x$ if $a_n = 1$, and $x_{\overline{a}(n-1)} > x$ if $a_n = -1$. Since $\overline{a}(m) \notin S$, and S is closed under restriction, if $b \in S$ and $|b| \geq m$, then $b_n = -a_n$ for some $n \leq m$; for that n, we have

$$|x - x_b| \geq 3^{-n}/2 \geq 3^{-m}/2,$$

by (2.2). On the other hand, if $b \in S$ and $|b| < m$, then $|y_b| > 3^{-m}/2$. Condition (2.3.1) now follows. $\square$

(2.4) Lemma *If B_0 is a countable subset of C, then there exists a detachable subset B of C such that*

(i) *each element of B has a restriction in B_0,*

(ii) *to each b in* B_0 *there corresponds* $n > |b|$ *such that each element of* C *of length n extending b has a restriction in B.*

In particular, B_0 *is a bar (respectively, uniform bar) for C if and only if B is a bar (respectively, uniform bar) for C.*

Proof. Let $c_1, c_2, \ldots$ be a one-one enumeration of C, and let $b_1, b_2, \ldots$ be an enumeration of B_0. Define

$$B \equiv \{c_i : c_i \text{ is an extension of } b_j \text{ for some } j \text{ in } \mathbb{N}^+ \text{ with } j \leq i\}.$$

Then B is detachable from C (we need the one-one enumeration of C at this point), and clearly satisfies (i). Given any $b \equiv b_j$ in B_0, choose $n > |b|$ so large that $i \geq j$ whenever $|c_i| \geq n$. If $|c_i| = n$ and c_i extends b, then, clearly, $c_i \in B$. Thus (ii) obtains.

It is clear from (i) and (ii) that B_0 is a bar if and only if B is a bar, and that if B is a uniform bar, then so is B_0. If B_0 is a uniform bar, choose m in $\mathbb{N}^+$ so that each path in C has a restriction in B_0 of length not exceeding m. Then, by (ii), for each a in C with $|a| = m$ there exists $n_a > m$ such that each extension of a in C with length n_a has a restriction in B. It follows that each path in C has a restriction in B with length not exceeding $\max\{n_a : a \in C, |a| = m\}$. Hence B is a uniform bar. □

Using the fan C, we introduce a special representation of certain compact intervals in $\mathbb{R}$.

(2.5) Lemma *If* $1/2 < \theta < 1$, *then for each infinite path a in C, the series* $\sum_{i=1}^{\infty} a_i\theta^i$ *converges to a point of the closed interval* $I \equiv [-\theta/(1-\theta), \theta/(1-\theta)]$; *and each point of I has the form* $\sum_{i=1}^{\infty} a_i\theta^i$ *for some path a in C.*

Proof. It is trivial that if a is an infinite path in C, then $\sum_{i=1}^{\infty} a_i\theta^i$ converges to a limit in I. Conversely, consider any x in I, set $a_0 \equiv 0$, and define $a_1, a_2, \ldots$ inductively so that for each $n \geq 1$,

$$a_n = -1 \Rightarrow x < \sum_{i=0}^{n-1} a_i\theta^i - \theta^n + \sum_{i=n+1}^{\infty} \theta^i$$

and

$$a_n = 1 \Rightarrow x > \sum_{i=0}^{n-1} a_i\theta^i + \theta^n - \sum_{i=n+1}^{\infty} \theta^i .$$

(Note that

$$\Sigma_{i=0}^{n-1} a_i\theta^i - \theta^n + \Sigma_{i=n+1}^{\infty} \theta^i - (\Sigma_{i=0}^{n-1} a_i\theta^i + \theta^n - \Sigma_{i=n+1}^{\infty} \theta^i)$$

$$= -2\theta^n + 2\Sigma_{i=n+1}^{\infty} \theta^i = -2\theta^n + 2\theta^{n+1}/(1-\theta)$$

$$= 2\theta^n(2\theta-1)/(1-\theta) > 0.)$$

Then $a \equiv (a_n)_{n=1}^{\infty}$ is a path in C. We prove by induction that

(2.5.1) $$|x - \Sigma_{i=0}^{n} a_i\theta^i| \leq \Sigma_{i=n+1}^{\infty} \theta^i$$

for $n = 0,1,2,\ldots$. This is trivial if $n = 0$. Suppose (2.5.1) holds for $n = k$, and consider, for example, the case where $a_{k+1} = 1$. Then

(2.5.2) $$x > \Sigma_{i=0}^{k+1} a_i\theta^i - \Sigma_{i=k+2}^{\infty} \theta^i.$$

By the induction hypothesis, we also have

(2.5.3) $$x \leq \Sigma_{i=0}^{k} a_i\theta^i + \Sigma_{i=k+1}^{\infty} \theta^i = \Sigma_{i=0}^{k+1} a_i\theta^i + \Sigma_{i=k+2}^{\infty} \theta^i.$$

Combining (2.5.2) and (2.5.3), we obtain (2.5.1) for $n = k+1$. This completes the induction. It follows that $\Sigma_{i=1}^{\infty} a_i\theta^i$ converges to x. □

The representation of I constructed in Lemma (2.5) is described by the diagram below, in which a path starting from the top node of the tree corresponds to a path in the fan C.

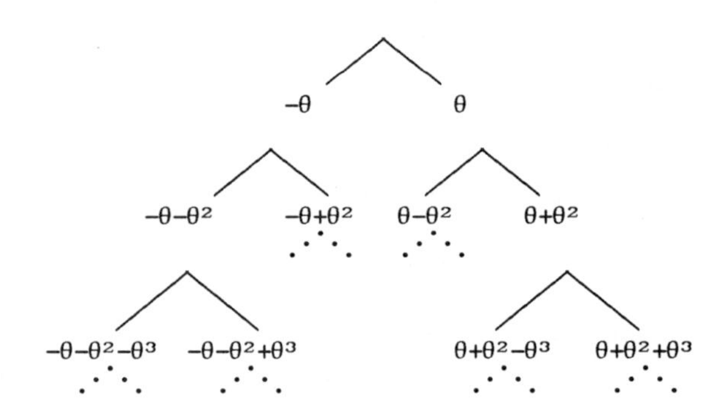

One more lemma will bring us to the key result in our present discussion. The last step of the proof of this lemma uses the fact that the image of a countable set is countable.

(2.6) Lemma *Let X be a countable set, and φ a map of X into $\mathbb{R}$. Then*

$$S \equiv \{x \in X : \varphi(x) > 0\}$$

is countable.

Proof. Let θ be a map of a detachable subset D of $\mathbb{N}$ onto X. For each $n \in \mathbb{N}$ construct a partition $A_n \cup B_n$ of $D \cap \{0,1,\ldots,n\}$ such that if $m \in A_n$, then $\varphi\theta(m) > 1/n$, and if $m \in B_n$, then $\varphi\theta(m) < 2/n$. Then $\cup_n A_n$ is a countable subset of $\mathbb{N}$ mapped by θ onto S; whence S is countable. □

(2.7) Theorem *If B is a detachable subset of C, then there exists a nonnegative uniformly continuous function f on $[0,1]$ such that*

(i) *$f(x) > 0$ for all x if and only if B is a bar for C,*

(ii) *$\inf f > 0$ if and only if B is a uniform bar for C.*

Conversely, if f is a nonnegative uniformly continuous mapping on $[0,1]$, then there exists a detachable subset B of C satisfying (i) and (ii).

Proof. Let B be a detachable subset of C, and define

$$S \equiv \{\alpha \in C : \text{no restriction of } \alpha \text{ is in } B\}.$$

Then S is a detachable subset of C that is closed under restriction. By (2.1), $P(S)$ is totally bounded, so that

$$f(x) \equiv \rho((x,0),P(S))$$

defines a nonnegative uniformly continuous function f on $[0,1]$. Consider any x in $[0,1]$, and construct an infinite path α in C as in (2.3). If B is a bar for C, then there exists m with $\overline{\alpha}(m) \notin S$, and therefore $f(x) \geq 3^{-m}/2$; moreover, if B is a uniform bar, then m can be chosen independent of x, so that $\inf f \geq 3^{-m}/2 > 0$. On the other hand, if f is everywhere positive, and α is any path in C, then as $f(\frac{1}{2} + \Sigma_{i=1}^{\infty} \alpha_i 3^{-i}) > 0$, we can find $m \geq 1$ such that

$$\rho((\tfrac{1}{2} + \Sigma_{i=1}^{m} \alpha_i 3^{-i}, 3^{-m}/2), P(S)) > 0.$$

Then $\overline{\alpha}(m) \notin S$, and therefore $\overline{\alpha}(n) \in B$ for some $n \leq m$; moreover, if $\inf f > 0$, then m can be chosen independent of α. This completes the proof that f satisfies (i) and (ii).

Conversely, let f be a nonnegative uniformly continuous function on $[0,1]$. Fix a rational number θ with $1/2 < \theta < 1$. Replacing f by the map

$$x \to f(\tfrac{1}{2}(\theta^{-1} - 1)x + \tfrac{1}{2})$$

on the closed interval $I \equiv [-\theta/(1-\theta), \theta/(1-\theta)]$, we may assume that f is defined on I. For each positive rational r, the set

$$K(r) \equiv \{(x,y) \in I \times I : |x-y| < r\}$$

is totally bounded, so that $M(r) \equiv \sup\{f(x) : x \in K(r)\}$ exists. Construct a map ϵ of the set $\mathbb{Q}^+$ of positive rationals into itself such that $M(r) < \epsilon(r) < M(r) + r$ for all r in $\mathbb{Q}^+$. Then as f is uniformly continuous, $\epsilon(r) \to 0$ as $r \to 0$. Now let

$$B_0 \equiv \{a \equiv (a_1,\ldots,a_n) \in C : \epsilon(\theta^{n+1}/(1-\theta)) < f(\Sigma_{i=1}^n a_i\theta^i)\}.$$

Then B_0 is countable, by (2.6). Construct a detachable subset B of C as in (2.4). If f is everywhere positive, and α is any infinite path in C, then as

$$0 < y \equiv f(\Sigma_{i=1}^\infty a_i\theta^i),$$

there exists n in $\mathbb{N}^+$ such that

$$f(\Sigma_{i=1}^n a_i\theta^i) > y/2 > \epsilon(\theta^{n+1}/(1-\theta)).$$

Thus $\overline{a}(n) \in B_0$, so that B_0, and therefore B, is a bar for C. Moreover, if $\inf f > 0$, then n can be chosen independent of α, so that B_0, and therefore B, is a uniform bar for C. On the other hand, if B is a bar for C, then so is B_0. In that case, if x is any point of I, and α is an infinite path in C with $x = \Sigma_{i=1}^\infty a_i\theta^i$, then, choosing a restriction $b \equiv (a_1,\ldots,a_n)$ of α in B_0, we have

$$|\Sigma_{i=1}^n a_i\theta^i - x| \le \Sigma_{i=n+1}^\infty \theta^i = \theta^{n+1}/(1-\theta),$$

so that, by our choice of $\epsilon(r)$ and the fact that $b \in B_0$,

$$f(x) > f(\Sigma_{i=1}^n a_i\theta^i) - \epsilon(\theta^{n+1}/(1-\theta)) > 0.$$

Moreover, if B, and therefore B_0, is a uniform bar, then b can be chosen from a certain finite subset F of B_0; whence, clearly, f is bounded away from 0, and therefore $\inf f > 0$. □

An immediate consequence of Theorem (2.7) is

(2.8) Corollary *The following statements are equivalent:*

(i) *every detachable bar of C is a uniform bar;*

(ii) *every uniformly continuous map of* [0,1] *into the positive real line has a positive infimum.* □

Thus, within BISH, statement (ii) is equivalent to the fan theorem.

From Theorem (2.7), and Proposition (3.1) of Chapter 5, we have

(2.9) Corollary *If CPF holds, then there exists a uniformly continuous map of* $[0,1]$ *into* $\mathbb{R}^+$ *with infimum* 0. □

Among the consequences of Corollary (2.9) are the following, whose proofs are relegated to the Problems section.

(2.10) Proposition *If CPF holds, then there is a compact subset of the open unit ball in* $\mathbb{R}^2$ *which has diameter* 2. □

Call a mapping f between metric spaces **injective** if $f(x) \neq f(y)$ whenever $x \neq y$.

(2.11) Proposition *If CPF holds, then there is an injective uniformly continuous map of the unit circle* $\{x \in \mathbb{R}^2 : \|x\| = 1\}$ *into* $\mathbb{R}^2$ *whose inverse is pointwise continuous but not uniformly continuous.* □

PROBLEMS

1. With reference to Lemma (2.1), give a Brouwerian example of a detachable subset S of C such that $P(S)$ is not totally bounded.
2. Assuming CPF, construct a countable bar B_0 for C that contains at most one sequence of each length. Construct B as in Lemma (2.4), and hence construct a detachable subset F of C such that

   (i) F is closed under restriction;

   (ii) each infinite path in C has a restriction that is not in F;

   (iii) F contains arbitrarily long finite sequences.
3. In the notation of Lemma (2.5), construct an infinite path $a \equiv (a_n)$ in C such that $\Sigma_{i=1}^{\infty} a_i\theta^i = \theta$.
4. Prove Proposition (2.10).
5. Assuming CPF, construct a pointwise continuous map f from a compact space into $\{0,1\}$, such that f is not uniformly continuous.
6. Prove Proposition (2.11).
7. A map f of a metric space X into a metric space X' is said to be **hyperinjective** if for each pair A,B of compact subsets of X with $\inf\{\rho(x,y) : x \in A, y \in B\} > 0$, there exists $r > 0$ such that $\rho(f(x),f(y)) \geq r$ whenever $x \in A$ and $y \in B$. Prove that if X is

compact, and f is both uniformly continuous and hyperinjective, then the inverse of f is uniformly continuous, and $f(X)$ is compact.

8. Assuming CPF, give an example of an injective uniformly continuous map that is not hyperinjective.

9. A metric space X is **stepwise connected** if for any a,b in X and $\epsilon > 0$, there exist points $x_0 = a, x_1, \ldots, x_n = b$ of X such that $\rho(x_{i-1},x_i) < \epsilon$ for $i = 1,\ldots,n$. Assuming CPF, construct a stepwise connected subset of $\mathbb{R}^2$ that is the union of two disjoint nonempty compact subsets.

NOTES

For a discussion of finitism, see the articles by Yessenin-Volpin in *Intuitionism and Proof Theory* (North Holland, 1970) and **Springer 873.**

Section 2 is based on *A uniformly continuous function on* [0,1] *that is everywhere different from its infimum*, by Julian and Richman (Pacific J. Math. 111 (1984), 333-340). For a different construction of such a function, see Aberth, *Computable Analysis* (McGraw-Hill, 1980, p. 70).

Chapter 7. Intuitionistic Logic and Topos Theory

In which an informal study of formal first-order intuitionistic logic is undertaken. In the first two sections, axioms for the propositional calculus and the predicate calculus are presented, and the use of Kripke models to demonstrate unprovability is illustrated. The last two sections contain two examples of topos models: one sheaf model and one presheaf model, the latter showing the unprovability of the world's simplest axiom of choice.

1. Intuitionistic propositional calculus

As we remarked in Chapter 1, the codification of intuitionistic logic, the logic of constructive mathematics, grows out of our mathematical experience. We do not develop the logic in the first instance, and then use that logic as a foundation for our mathematics; rather, we formulate our rules of logic to reflect our mathematical practice. Nevertheless, the abstraction of logical axioms clarifies much of the mathematical experience on which it is based.

Propositional calculus is a formal system for studying the connectives $\vee$ (or), $\wedge$ (and), $\Rightarrow$ (implies), and $\neg$ (not), which can be used to form new sentences, or propositions, from old ones. The basic ingredients of the propositional calculus are these four connectives and an infinite sequence $p,q,r,\ldots$ of **propositional variables.** These symbols may be combined, with parentheses, to form **well-formed formulae** such as

$$(p \Rightarrow (q \vee r)) \wedge \neg q,$$

which may be thought of as **propositional functions:** if we replace p,q and r by propositions, we get another proposition. Certain well-formed formulae, such as $((p \vee q) \Rightarrow (q \vee p))$ and $\neg(p \wedge \neg p)$, give true sentences no matter what sentences are substituted for the variables; these are called **tautologies.** Tautologies are the patterns of sentences that are true on the basis of their logical structure alone. The study of the propositional calculus revolves around describing the tautologies.

The proof-theoretic approach to describing tautologies is to write down a list of basic tautologies, called **axioms**, and to specify certain **rules of inference**, which generate tautologies from other tautologies. If this is done correctly, then the tautologies will be exactly those well-formed formulae that can be generated from the axioms using the rules of inference. The process of generating a formula from the axioms is called a **proof**, or a **derivation**, of that formula, and such a formula is called a **theorem.** Standard axioms for the **intuitionistic propositional calculus** are as follows, where A, B and C are arbitrary well formed formulae.

(1.1)
1. $A \Rightarrow (A \wedge A)$
2. $(A \wedge B) \Rightarrow (B \wedge A)$
3. $(A \Rightarrow B) \Rightarrow (A \wedge C \Rightarrow B \wedge C)$
4. $(A \Rightarrow B) \Rightarrow ((B \Rightarrow C) \Rightarrow (A \Rightarrow C))$
5. $B \Rightarrow (A \Rightarrow B)$
6. $(A \wedge (A \Rightarrow B)) \Rightarrow B$
7. $A \Rightarrow (A \vee B)$
8. $(A \vee B) \Rightarrow (B \vee A)$
9. $((A \Rightarrow C) \wedge (B \Rightarrow C)) \Rightarrow (A \vee B \Rightarrow C)$
10. $\neg A \Rightarrow (A \Rightarrow B)$
11. $((A \Rightarrow B) \wedge (A \Rightarrow \neg B)) \Rightarrow \neg A$

The only rule of inference needed is **modus ponens:** from A and $A \Rightarrow B$, infer B - that is, the formulae A and $A \Rightarrow B$ together generate the formula B. We get axioms for the *classical* propositional calculus by adding to the above list the **law of excluded middle**

EM $$A \vee \neg A.$$

On the other hand, the model-theoretic approach to identifying tautologies concentrates on the meaning of the connectives, rather than on the mechanics of proof. Given a well-formed formula in the classical propositional calculus, we substitute zeroes and ones for its variables, and reduce according to the usual truth tables of the connectives:

$\Rightarrow$	0	1
0	1	1
1	0	1

$\vee$	0	1
0	0	1
1	1	1

$\wedge$	0	1
0	0	0
1	0	1

$\neg$	0	1
	1	1

More precisely, a **classical model** for the propositional calculus is a function α from the set of propositional variables to $\{0,1\}$, which is extended to a function on the set of all well-formed formulae by means of the following rules:

$$\alpha(A \vee B) = \alpha(A) + \alpha(B) - \alpha(A)\alpha(B),$$
$$\alpha(A \wedge B) = \alpha(A)\alpha(B),$$
$$\alpha(\neg A) = 1 - \alpha(A),$$
$$\alpha(A \Rightarrow B) = 1 - \alpha(A)(1 - \alpha(B)).$$

A well-formed formula A **holds** in the model α if $\alpha(A) = 1$. It is a standard result that the theorems of the classical propositional calculus are exactly those well-formed formulae that hold in all classical models.

There is an analogous result for the intuitionistic propositional calculus. A (finite) **Kripke model** $\mathcal{M}$ for the intuitionistic propositional calculus is a finite set of classical models, which are also called **states**, partially ordered by setting $\alpha \leq \beta$ if $\alpha(p) \leq \beta(p)$ for each propositional variable p. Each α in $\mathcal{M}$ is extended inductively to a function on the set of all well-formed formulae as follows:

$$\alpha(A \vee B) = \alpha(A) + \alpha(B) - \alpha(A)\alpha(B),$$
$$\alpha(A \wedge B) = \alpha(A)\alpha(B),$$
$$\alpha(\neg A) = 1 - \max\{\beta(A) : \beta \geq \alpha\},$$
$$\alpha(A \Rightarrow B) = 1 - \max\{\beta(A)(1 - \beta(B)) : \beta \geq \alpha\}.$$

We say that a well-formed formula A **holds** in a Kripke model $\mathcal{M}$ if $\alpha(A) = 1$ for each α in $\mathcal{M}$. It turns out that a well-formed formula is a theorem in the intuitionistic propositional calculus if and only if it holds in every Kripke model.

We may think of sets of states of a Kripke model as being **truth values**: the truth value of a formula A is the set $\|A\|$ of states α such that $\alpha(A) = 1$. The truth value of a formula A is a **filter**: that is, if $\alpha \in \|A\|$, and $\alpha \leq \beta$, then $\beta \in \|A\|$. Note that $\|A \vee B\| = \|A\| \cup \|B\|$; that $\|A \wedge B\| = \|A\| \cap \|B\|$; that $\|A \Rightarrow B\|$ is the largest filter whose intersection with $\|A\|$ is contained in $\|B\|$; and that $\|\neg A\|$ is the largest filter whose intersection with $\|A\|$ is empty. A formula A holds in the model when $\|A\|$ is the set of all states of the model; thus truth corresponds to the set of all states.

Proofs and models play complementary roles: to establish that a formula is a theorem, we construct a proof; to establish that a formula is not a theorem, we exhibit a model in which it does not hold. For example, to see that $p \vee \neg p$ is not a theorem, consider a Kripke model $\mathcal{M} \equiv \{\alpha,\beta\}$ such that $\alpha \leq \beta$, $\alpha(p) = 0$ and $\beta(p) = 1$. Then $\alpha(\neg p) = 0$, because $\beta(p) = 1$; so $\alpha(p \vee \neg p) = 0$, and therefore $p \vee \neg p$ does not hold in $\mathcal{M}$.

In the last example, we may think of the classical models α and β as two states of knowledge of the creating subject, occurring at times t_α and t_β with $t_\alpha < t_\beta$. The sentence p has not been established at time t_α, so it cannot be said to hold then. On the other hand, we cannot assert $\neg p$ at time t_α, as p itself will be established at the later time t_β. Thus at t_α we can neither assert p nor assert $\neg p$, so $p \vee \neg p$ does not hold then.

In general, we should not expect the states of a Kripke model to be linearly ordered, as the creating subject may have several alternative courses of action leading from a state, and states attained by different courses of action may be incompatible.

2. Predicate calculus

To describe the finer structure of sentences, we introduce predicates and quantifiers. We dispense with variables representing propositions, and introduce variables representing elements of whatever mathematical structure we are investigating. The resulting formal system is called a **predicate calculus.**

A **first-order language** consists of the connectives $\wedge$, $\vee$, $\neg$, $\Rightarrow$, a list of variables and constants, a list of predicate symbols, and the quantifiers $\forall$ (for each) and $\exists$ (there exists). Each predicate symbol has a positive integer associated with it indicating how many places it has. If P is an n-place predicate, and $a_1,\ldots,a_n$ are variables or constants, then $P(a_1,\ldots,a_n)$ is a well-formed formula. From these atomic well-formed formulae we can construct others by means of the connectives and by applying the quantifiers $\exists$ and $\forall$ according to the scheme: if A is a well-formed formula, and x is a variable, then $\exists xA$ and $\forall xA$ are well-formed formulae.

In order to formulate a set of axioms for the predicate calculus, we must introduce some technical notation and terminology. We denote by $A(x/t)$ the formula obtained on replacing every occurrence of the variable x in A by t, which can be either a variable or a constant. An occurrence of the variable x in a formula A is **bound** if it appears in a subformula of the form $\forall xB$ or $\exists xB$; otherwise the occurrence is said to be **free**. Let x be a variable, t a variable or a constant, and A a formula; we say that **t is free for x in A** if no free occurrence of x in A is in a subformula of A of the form $\forall tB$.

The axioms of the **intuitionistic predicate calculus** are obtained by adding to the propositional axioms (1.1) those in the following list; we also add the rule of inference called **generalisation**, which allows us to infer $\forall xA$ from A.

(2.1)
1. $\forall x(A \Rightarrow B) \Rightarrow (A \Rightarrow \forall xB)$ if x is not free in A
2. $\forall x(A \Rightarrow B) \Rightarrow (\exists xA \Rightarrow B)$ if x is not free in B
3. $\forall xA \Rightarrow A(x/t)$ if t is free for x in A
4. $A(x/t) \Rightarrow \exists xA$ if t is free for x in A

A **classical model** $\mathcal{M}$ for a first-order language $\mathcal{L}$ is a nonempty set S together with an assignment V of an element of S to each constant in $\mathcal{L}$, and a subset of S^n to each n-place predicate symbol of $\mathcal{L}$. To define what it means for a sentence to **hold** in $\mathcal{M}$, we first enlarge $\mathcal{L}$ so that the elements of S are constants of $\mathcal{L}$; without loss of generality we may assume that S is the set of constants of $\mathcal{L}$, and that V is the identity on S. We then proceed inductively as follows:

(2.2)
1. If P is an n-place predicate symbol, and $c_1,\ldots,c_n$ are constants, then $P(c_1,\ldots,c_n)$ holds in $\mathcal{M}$ if and only if $(c_1,\ldots,c_n) \in V(P)$.
2. $A \vee B$ holds in $\mathcal{M}$ if and only if A holds in $\mathcal{M}$ or B holds in $\mathcal{M}$.
3. $A \wedge B$ holds in $\mathcal{M}$ if and only if A holds in $\mathcal{M}$ and B holds in $\mathcal{M}$.
4. $\neg A$ holds in $\mathcal{M}$ if and only if A does not hold in $\mathcal{M}$.
5. $A \Rightarrow B$ holds in $\mathcal{M}$ if and only if B holds in $\mathcal{M}$ whenever A holds in $\mathcal{M}$.
6. $\exists xA$ holds in $\mathcal{M}$ if and only if there is a constant c such that $A(x/c)$ holds in $\mathcal{M}$.
7. $\forall xA$ holds in $\mathcal{M}$ if and only if $A(x/c)$ holds in $\mathcal{M}$ for each

constant c.

A **Kripke model** $\mathcal{M}$ for a first-order language $\mathcal{L}$ consists of

(i) a nonvoid partially ordered set of **states**;

(ii) a mapping which assigns to each state α a classical model $\mathcal{M}_\alpha \equiv (S_\alpha, V_\alpha)$ of $\mathcal{L}$, such that if α,β are states with $\alpha \leq \beta$, then $S_\alpha \subset S_\beta$, and $V_\alpha(P) \subset V_\beta(P)$ for each predicate symbol P.

Given a Kripke model $\mathcal{M}$, we want a formal definition which reflects the intuitive interpretation of the notation

$$\mathcal{M} \Vdash_\alpha A$$

as 'the creating subject knows at stage α that A is true'. For example, on this interpretation we want $\mathcal{M} \Vdash_\alpha (A \Rightarrow B)$ to hold if, whenever the creating subject has a proof of A at some stage $\beta \geq \alpha$, he has a proof of B at stage β. Similarly, we want $\mathcal{M} \Vdash_\alpha \forall x A$ to hold precisely when $A(x/c)$ holds for any constant c that is already, or will be, constructed by the creating subject. The requirements of these two examples are captured by parts 4 and 7 of the definition below.

Let α be a state of $\mathcal{M}$, and A,B sentences in the language $\mathcal{L}$ augmented by the constants in S_α. Define $\mathcal{M} \Vdash_\alpha A$ inductively as follows:

(2.3) 1. If A is atomic, then $\mathcal{M} \Vdash_\alpha A$ if and only if A holds in $\mathcal{M}_\alpha$.

2. $\mathcal{M} \Vdash_\alpha A \vee B$ if and only if either $\mathcal{M} \Vdash_\alpha A$ or $\mathcal{M} \Vdash_\alpha B$.

3. $\mathcal{M} \Vdash_\alpha A \wedge B$ if and only if $\mathcal{M} \Vdash_\alpha A$ and $\mathcal{M} \Vdash_\alpha B$.

4. $\mathcal{M} \Vdash_\alpha (A \Rightarrow B)$ if and only if $\mathcal{M} \Vdash_\beta B$ whenever $\alpha \leq \beta$ and $\mathcal{M} \Vdash_\beta A$.

5. $\mathcal{M} \Vdash_\alpha \neg A$ if and only if $\mathcal{M} \Vdash_\beta A$ does not hold for any $\beta \geq \alpha$.

6. $\mathcal{M} \Vdash_\alpha \exists x A$ if and only if $\mathcal{M} \Vdash_\alpha A(x/c)$ for some c in S_α.

7. $\mathcal{M} \Vdash_\alpha \forall x A$ if and only if $\mathcal{M} \Vdash_\beta A(x/c)$ whenever $\alpha \leq \beta$ and $c \in S_\beta$.

If $\mathcal{M} \Vdash_\alpha A$, we say that **A holds at α** in the given Kripke model. One can show that

If A is a sentence in $\mathcal{L}$, and there exists a state α of a Kripke model such that A does not hold at α, then A is not derivable in the intuitionistic predicate calculus.

The above result is the basis of many intuitionistic counterexamples to theorems of classical first-order logic. Consider, for example, the classical theorem

$$\forall x(P(x) \vee \forall yQ(y)) \Rightarrow (\forall xP(x) \vee \forall yQ(y)),$$

where P,Q are one-place predicate symbols, and x,y are distinct variables. Construct a Kripke model $\mathcal{M}$ as follows:

$$\text{the states of } \mathcal{M} \text{ are } 0 \text{ and } 1, \text{ with } 0 \leq 1;$$
$$S_0 = \{0\} \text{ and } S_1 = \{0,1\};$$
$$V_0(P) = \{0\} \text{ and } V_0(Q) = \phi;$$
$$V_1(P) = \{0\} \text{ and } V_1(Q) = \{0,1\}.$$

By (2.3,7), (2.3,1), and (2.2), we have $\mathcal{M} \Vdash_1 \forall yQ(y)$; so $\mathcal{M} \Vdash_1 P(0) \vee \forall yQ(y)$ and $\mathcal{M} \Vdash_1 P(1) \vee \forall yQ(y)$, by (2.3,2); whence

$$\mathcal{M} \Vdash_1 \forall x(P(x) \vee \forall yQ(y)),$$

by (2.3,7). Also $\mathcal{M} \Vdash_0 P(0) \vee \forall yQ(y)$, by (2.3,1) and (2.2); so $\mathcal{M} \Vdash_0 \forall x(P(x) \vee \forall yQ(y))$. But, in view of (3.2,1), $Q(0)$ does not hold at 0, and $P(1)$ does not hold at 1; so $\forall xP(x) \vee \forall yQ(y)$ does not hold at 0.

If the set of states of a Kripke model $\mathcal{M}$ is countable, and $\{\alpha : \alpha \leq \beta\}$ is finite for each state β, then we can use the standard graph theoretical representation of such a partially ordered set as a shorthand description of $\mathcal{M}$, by decorating each node α with the sets S_α and $V_\alpha(P)$. Because $S_\alpha \subset S_\beta$ and $V_\alpha(P) \subset V_\beta(P)$ whenever $\alpha \leq \beta$, it suffices to decorate the node α with those elements of S_α that do not appear previously, and with those formulae $P(c_1,\ldots,c_n)$ that hold at, and not before, β. We shall draw such graphs from the top down, so that $\alpha \leq \beta$ if and only if the node for α is joined to, and lies above, the node for β. Thus the **Kripke tree** for the Kripke model constructed in the example above is

```
0 | P(0)
  |
1 | Q(0),Q(1)
```

For another example, consider Markov's principle

$$(\forall x(P(x) \vee \neg P(x)) \wedge \neg\forall x\neg P(x)) \Rightarrow \exists xP(x)$$

in **Heyting arithmetic** – that is, the intuitionistic version of first-order (Peano) arithmetic. Here P is a one-place predicate symbol, and x is a variable running over the natural numbers. A model in which Markov's principle fails is given by the infinite Kripke tree below:

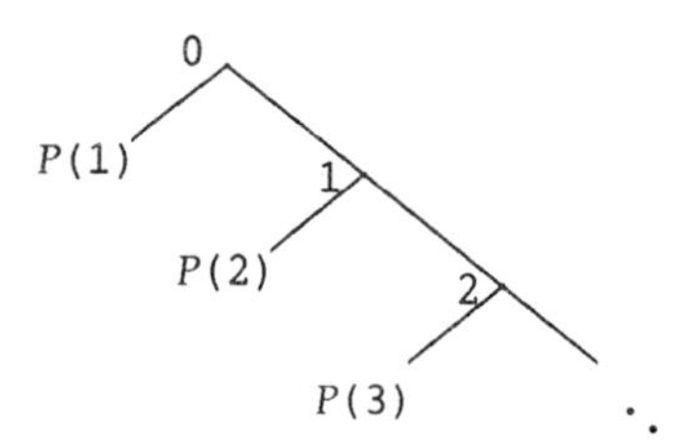

3. The sheaf model C(X)

We think of constructive proofs as referring to the familiar denizens of the mathematical universe, with special care being taken over what counts as a proof of existence. But our arguments may be construed to refer to a totally different world: a topos model. Although at first we may not care that our constructive theorems, quite unintentionally, also hold in this exotic setting, topos models provide a source of counterexamples that can play the same role as Brouwerian and recursive counterexamples: *if there is a topos model in which a theorem T does not hold, then T does not admit a constructive proof.*

A **topos** is a category that behaves like the category of sets. We can identify within a topos objects that play the role of various standard sets: for example, the empty set, one-element sets, the set of natural numbers, the set of real numbers.

We have to be rather careful when dealing with topos models, since dependent choice does not always hold in a topos. If we allow the axiom of dependent choice, which we normally do, we must check that it holds in the model in question.

The first topos that we will consider is the category of sheaves over a topological space X. Although countable choice need not hold in such a model - so its utility as a source of counterexamples is diminished - it is relatively easy to describe how statements about real numbers are interpreted therein.

Let X be a (classical) topological space and $C(X)$ the space of continuous real-valued functions on X. We think of the elements of $C(X)$ as if they were real numbers, and we interpret statements about real numbers as statements about elements of $C(X)$. In this interpretation, the rational number q is identified with the constant function $q \in C(X)$ defined by $q(x) \equiv q$.

If $x \in X$, $f \in C(X)$, and P is a statement about a real number, then P is said to be **true for f at x** if P is a true statement about the real number $f(y)$ for all y in some neighbourhood of x in X. If P is true for f at each x in X, we say that P is **true for f**; if there is no point x in X at which P is true for f, we say that P is **false for f**. For example, if P is the statement $f > 0$ in $C(X)$, then P is true for f at all points x of X which have a neighbourhood U such that $f(y) > 0$ for all y in U; and P is false for f at all other points of X.

In general, a statement P need not be either true or false for a given f in $C(X)$; rather, there will be some open subset of X where P is true for f. Thus the **truth values** in this model are open subsets of X, with the subset X representing truth, and the empty subset falsehood.

The constructive version of trichotomy holds in such models: that is, if $f < g$ is true, then so is ($f < h$ or $h < g$). Indeed, if $f < g$ and $x \in X$, then the inequality $f(x) < g(x)$ entails either $f(x) < h(x)$ or $h(x) < g(x)$; since f, g, and h are continuous, there is a neighborhood N of x such that either $f(y) < h(y)$ for each y in N, or $h(y) < g(y)$ for each y in N; as x is arbitrary, we conclude that ($f < h$ or $h < g$) is true. But the classical version of trichotomy need not hold. To see this, let X be the unit interval $[0,1]$, and consider $f(x) \equiv x$ and $g(x) \equiv 0$: the truth value of $f \leq g$ is the empty set, while the truth value of $f > g$ is the set $(0,1]$; so the truth value of the statement ($f \leq g$ or $f > g$) is $(0,1]$, which is not equal to X. Since we have a topos model in which the classical law of trichotomy does not hold, we conclude that that law is not provable within BISH.

Similarly, the classically true statement

if $x,y \in \mathbb{R}$ and $xy = 0$, then either $x = 0$ or $y = 0$

cannot be proved within BISH. To see this, take $X \equiv \mathbb{R}$, $f(x) \equiv \max\{0,x\}$, and $g(x) \equiv \max\{0,-x\}$; then the truth value of ($fg = 0$) is X, but that of ($f = 0$ or $g = 0$) is $(-\infty,0) \cup (0,\infty)$.

In $C(X)$ each real number can be approximated arbitrarily closely by rational numbers, at least locally (and local truth is what counts). Consider the real number $f(x) \equiv x$, where $X \equiv [0,1]$. The statement

$$|f| < 1/n \text{ or } |f - 1/n| < 1/n \text{ or } |f - 2/n| < 1/n \text{ or } \cdots \text{ or } |f - 1| < 1/n$$

is true for f at each x in X. On the other hand, we cannot construct a sequence of rationals converging to f, because if a real number g in $C(X)$ is the limit of a sequence of rational numbers, then g is locally constant: each x is contained in a neighborhood U for which there exists a sequence of rational numbers (q_n) such that $q_n \to g(y)$ for each y in U. Thus we must distinguish between **Cauchy reals**, which are limits of Cauchy sequences of rational numbers, and **Dedekind reals**, which can be approximated arbitrarily closely by rational numbers. Note that countable choice fails in this model.

Each real number in $C(X)$ determines a **Dedekind cut:** that is, a a pair of disjoint nonempty sets of rational numbers L and U such that

(i) every rational less than an element of L is in L,

(ii) every rational greater than an element of U is in U,

(iii) if p,q are rationals with $p < q$, then either $p \in L$ or $q \in U$.

As in the classical context, a set S of rationals can be identified with its **characteristic function** χ_S, which maps each rational number $q \in C(X)$ to the truth value of the statement $(q \in S)$. In this case, the truth values are open sets, and U assigns to the rational number q the set $\{x \in X : f(x) < q\}$.

Consider the relationship between the statement $f \neq 0$ and $f = 0$. The truth value of $f \neq 0$ is the open set $\{x \in X : f(x) \neq 0\}$. The truth value of $\neg(f \neq 0)$ is the interior of $\{x \in X : f(x) = 0\}$, which is the same as the truth value of $f = 0$. Thus $\neg(f \neq 0)$ is equivalent to $f = 0$; this equivalence is a constructively valid theorem about the real numbers. On the other hand, since every real number z which satisfies $\neg(z = 0)$ is the limit of a sequence (z_n) of real numbers distinct from 0, the truth value of $\neg(f = 0)$ is the interior of the closure of $\{x \in X : f(x) \neq 0\}$; in general, this set will be larger than the truth value of $f \neq 0$. Thus $f \neq 0$ implies $\neg(f = 0)$, as expected, but the converse does not hold.

4. Presheaf topos models

In this final section of the book we shall illustrate the use of presheaf topos models by constructing one in which the world's simplest axiom of choice fails. Our presentation will be very informal, and we will assume some familiarity with elementary category theory.

A **presheaf topos** is constructed by taking a small category $\mathcal{C}$ (ours will be finite) and considering the category $\mathcal{T}$ of all functors from $\mathcal{C}$ to the category of sets.

We first show how a presheaf topos can be considered as a generalisation of a Kripke model. Any partially ordered set becomes a category if we postulate a unique morphism from u to v exactly when $u \leq v$; a Kripke model for a first-order language $\mathcal{L}$ may then be thought of as a functor from a partially ordered set to the set of classical models of $\mathcal{L}$. It is often convenient to expand the notion of a Kripke model to allow arbitrary maps from S_α to S_β, rather than just the inclusion of S_α in S_β (we are using here the notation of the definition of 'Kripke model' in Section 2). If we expand a little more by replacing the underlying partially ordered set by a category, we get the notion of a **presheaf topos model.**

The name 'presheaf' comes from the case where $\mathcal{C}$ is the set of open sets in a topological space, viewed as a partially ordered set by defining $U \leq V$ to mean $V \subset U$.

For our illustration we take $\mathcal{C}$ to be a category with two objects and four morphisms: the two identity maps, one map between the objects, and one map from the target object to itself whose square is the identity. An **object** in the corresponding category $\mathcal{T}$ is a pair (S,σ), where $S \equiv (S_1,S_2)$ is a pair of sets, and $\sigma \equiv (\sigma_1,\sigma_2)$ is a pair of maps $\sigma_1:S_1 \to S_2$ and $\sigma_2:S_2 \to S_2$ such that $\sigma_2\sigma_1 = \sigma_1$ and σ_2^2 is the identity map on S_2; for convenience, we will often denote this object simply by S. We may think of S_1 and S_2 as present and future states of the set under consideration, and σ_2 as a symmetry of the future structure of the set, causing x and $\sigma_2 x$ to be indistinguishable in advance.

The following example will be important in our illustration. Let $M \equiv (M_1,M_2)$, where M_1 is the empty set and $M_2 \equiv \{0,1\}$; and let μ_1 be the inclusion of M_1 in M_2, and μ_2 the nontrivial permutation on M_2. Then the element (M,μ) of $\mathcal{T}$ can be thought of as a set that at present contains no elements, but in the future will contain two elements; although in the future we will be able to recognise 0 and 1 as distinct elements of our set, there is no way we can do so in advance.

To turn $\mathcal{T}$ into a category, it remains to define the morphisms between objects of $\mathcal{T}$; we do this in the usual way for categories of functors. Thus a **morphism** f between two objects (A,α) and (B,β) of $\mathcal{T}$ is a pair of functions $f_1:A_1 \to B_1$ and $f_2:A_2 \to B_2$ such that $\beta_1 f_1 = f_2\alpha_1$ and $\beta_2 f_2 = f_2\alpha_2$; that is, a morphism in $\mathcal{T}$ is a **natural transformation of functors** from $\mathcal{C}$ to the category of sets. The set of morphisms between objects A and B will be denoted by $[A,B]$. Each object in $\mathcal{T}$ admits a unique morphism into the object $\mathbf{1} \equiv (\{0\},\{0\})$ – that is, $\mathbf{1}$ is a **terminal object** of $\mathcal{T}$. Since the terminal objects in the category of sets are precisely the one-element sets, $\mathbf{1}$ plays the role of a one-element set in $\mathcal{T}$.

The categorical product $A \times B$ in $\mathcal{T}$ has the cartesian products $A_1 \times B_1$ and $A_2 \times B_2$ as its two sets, and has the obvious maps. This categorical product plays the role of the cartesian product in $\mathcal{T}$. To identify the object in $\mathcal{T}$ that plays the role of the set of all functions from A to B (the set $[A,B]$ is not an object of $\mathcal{T}$), we seek an object B^A that has the characteristic property

(4.1) $$[X \times A,B] \cong [X,B^A]$$

for all objects X, where $\cong$ denotes natural equivalence of functors. In the category of sets, the set B^A of all functions from A to B has the property (4.1).

It is straightforward to check that the pair (F,φ), defined as follows, satisfies the characteristic property of B^A in the category $\mathcal{T}$:

F_1 is the set $[A,B]$ of natural transformations from A to B,
F_2 is the set of all functions from A_2 to B_2,
$\varphi_1:F_1 \to F_2$ takes (f_1,f_2) to f_2,
$\varphi_2:F_2 \to F_2$ takes f to $\beta_2 f\alpha_2$.

Note that the image of $\varphi_1 F_1$ consists of those elements in F_2 that are invariant under φ_2.

We must distinguish carefully between the set $[A,B]$ and the object B^A; the latter is what is referred to when we interpret statements like 'there exists a function from A to B such that...' in $\mathcal{T}$. For example, with M the object defined above, $[\mathbf{1},M]$ is empty, because there are no functions from 1_1 to $M_1 = \phi$; but M satisfies the characteristic property of M^1, because $X \times 1$ is isomorphic to X.

Each topos comes equipped with certain truth values, which may be identified with the subobjects of the terminal object **1**. In the classical category of sets there are two subsets of $\mathbf{1} \equiv \{0\}$, hence two truth values. In the constructive category of sets there are many subsets of $\{0\}$, such as $\{n : n = 0$ and there is an odd perfect number$\}$, that cannot be identified with $\{0\}$ or ϕ. In our topos $\mathcal{T}$ there are three truth values, which may be thought of as 'true', 'false', and 'false now but true later'. These truth values are assigned to statements about objects of $\mathcal{T}$ by a straightforward generalisation of (2.3). Thus, for example, the statement 'there is a function from **1** to M' is not true in $\mathcal{T}$, as $(M^1)_1$ is empty; nor is the statement false, because $(M^1)_2$ is nonempty.

At last we come to our illustration of the use of presheaf topos models. Define N by $N_1 \equiv \phi$ and $N_2 \equiv \{\{0,1\}\}$, and let $F \equiv M^N$. Then F_1 is empty, and $F_2 \cong M_2 = \{0,1\}$. Thus the statement 'there is a map from N to M' – that is, 'there is an element of F' – is neither true nor false in $\mathcal{T}$. But the statement 'N is a set of two-element sets and contains at most one element' is true in $\mathcal{T}$, as is the statement 'M is the union of N'. Therefore the world's simplest axiom of choice is not true in $\mathcal{T}$. It follows that that axiom cannot be proved within the intuitionistic predicate calculus.

PROBLEMS

1. Prove that if A is a sentence, then $\neg\neg A \Rightarrow A$ cannot be derived in the intuitionistic propositional calculus.

2. Use appropriate Kripke models to refute each of the following statements about a sentence A:

   (i) $\neg\neg A \vee \neg A$

   (ii) $(\neg\neg A \Rightarrow A) \Rightarrow (A \vee \neg A)$.

3. Prove that if P is a one-place predicate symbol, and x is a variable running over the natural numbers, then the statements $\neg\neg\forall x(P(x) \vee \neg P(x))$ and $\neg\neg\forall x(\neg\neg P(x) \Rightarrow P(x))$ cannot be derived in Heyting arithmetic.

4. For each of the following formulae find a refutation by a Kripke model, where A and B are sentences, and P is a one-place predicate symbol.

   (i) $\neg(A \wedge B) \Rightarrow \neg A \vee \neg B$

   (ii) $\exists x(A \vee P(x)) \Rightarrow A \vee \exists x P(x)$

   (iii) $(A \Rightarrow \exists x P(x)) \Rightarrow \exists x(A \Rightarrow P(x))$

   (iv) $\neg\neg\exists x P(x) \Rightarrow \exists x \neg\neg P(x)$

   (v) $\forall x \neg\neg P(x) \Rightarrow \neg\neg\forall x P(x)$

5. Use a sheaf model $C(X)$ to show that neither of the following theorems about real numbers admits a constructive proof.

   (i) $\neg(x = y) \Rightarrow x \neq y$

   (ii) $x \leq y$ or $y \leq x$

6. Prove that the explicit description of B^A in the topos $\mathcal{T}$ of Section 4 produces an object of $\mathcal{T}$ that satisfies (4.1)

NOTES

Details of classical and intuitionistic logic may be found in *Mathematical logic, a first course* (Benjamin, 1969), by Joel W. Robbin.

Axioms (the same as ours) for the intuitionistic propositional calculus first appeared in A. Heyting, *Die formalen Regeln der intuitionistischen Logik* (Sitzungsber. preuss. Akad. Wiss. Berlin (1930), 42–56); see also pages 105–106 of Heyting's *Intuitionism, an Introduction* (3rd edn., North-Holland, 1971). The axioms for the predicate calculus are also standard, and can be found in Robbin's book and on page 980 of Troelstra's article *Aspects of constructive mathematics*, in *Handbook of Mathematical Logic* (J. Barwise ed., North-Holland, 1978).

In the context of the study of intuitionistic logic, it is worth noting the following comment made by Stolzenberg in his review of **Bishop** (Bull. Amer. Math. Soc. 76 (1970), 301–323):

> *This is as good a place as any to say – if it needs saying – that defining formal systems, constructively, and proving theorems about them, constructively, is a part of constructive mathematics. This is so regardless of the constructive content, or lack of it, in the ideas*

which the system is designed to formalize.

Our treatment of topos models is based on the work of Andre Scedrov. In *Diagonalization of continuous matrices as a representation of the intuitionistic reals* (Ann. Pure and Applied Logic 30 (1986)), Scedrov applies constructive theorems about real numbers to the space $C(X)$. See also his paper *Some aspects of categorical semantics: sheaves and glueing*, in *Logic Colloquium* '85 (ed. by M. A. Dickmann et al., North-Holland, 1986). The presheaf topos counterexample to the world's simplest axiom of choice is from Fourman and Scedrov, The *'world's simplest axiom of choice'* fails (Manuscripta Math. 38 (1982), 325-332).

For a highly readable exposition of elementary topos theory, see *Topoi*, by R. Goldblatt (North-Holland, 1979).

Index